CRYPTONOMICS

A MODERN INVESTORS HANDBOOK FOR CRYPTOCURRENCY MARKETS AND BLOCKCHAIN LITERACY

CHRIS WISER

Legal Disclaimer

The content contained in this book is intended for use as educational material only. While every effort is made to keep the information accurate and up to date, unintended errors may occur, especially due to the immature and ever-changing nature of the subject matter. All opinions expressed are formed by the personal research and experience of the author and are intended for informational purposes only. No material contained herein represents an endorsement of, or solicitation to invest in, any entity or asset mentioned. The author is not a registered investment, legal, or tax advisor. Before investing, you should always take your personal circumstances into account, conduct your own research, and consult a registered investment advisor. All investment decisions and trading strategies are made at your own risk. The author assumes no liability for loss or damage you may incur.

Contents

Introduction

2017 AND 2018 were eventful years for cryptocurrency.

We saw ordinary people become millionaires, and some even billionaires, due to this esoteric digital money called Bitcoin; we saw some of the largest companies and governments update their posture from dismissive to supportive; and we saw cryptocurrency land on the radar of mainstream news channels and investors.

These changes sparked widespread interest and produced new supporters, opponents, regulations, challenges, and opportunities. This undeniable progress invited an influx of new entrants seeking to understand or profit from this innovation. Today, many well known governments, multinational corporations, and financial institutions are looking at getting in on the action.

But what is Bitcoin? What is a cryptocurrency or altcoin? Where did they come from and who created them? Are they safe? Are we sure? What backs or underpins them?

Even if you are persistent enough to answer these questions by navigating the misinformation and incompleteness that plagues online forums and sporadic news articles, there is still the barrier of knowing how to invest in this new asset class in an intelligent, prudent way.

Is it a currency or investment? How do I navigate exchanges safely? How do I evaluate these cryptocurrencies for investment potential?

While the relatively high barrier to *fully* understanding cryptocurrency has deterred some, others have thrown caution to the wind on "investing" without conducting sufficient due diligence or respecting the risks involved.

Throughout history, radical innovations have often been greeted with this uniquely passionate and dichotomous following until they either fail or reach mainstream adoption.

Meanwhile, this deeply polarized ecosystem allows emotion to win out, leaving little content available for self-education and common sense analysis.

How is an interested investor supposed to learn in this environment?

There is a certain levelheadedness that needs to be brought to the conversation, and this book promises to do that.

In the process of writing this book, I have met far too many people who have attempted to become literate in cryptocurrency but, instead, became jaded by the "veterans" that were more interested in spouting jargon than helping educate. Therefore, this book promises to be a no-nonsense primer on cryptocurrency, a how-to-guide for investing in it successfully, and an unsensationalized and impartial analysis of its promise—both in the present and near future.

Together, we will explore the entire cryptocurrency industry from top to bottom. More importantly, you will be given the tools and resources needed to capitalize on the good this market has to offer, while avoiding the pitfalls that have broken many aspiring investors.

As an active early adopter, I've thoroughly experienced the cryptocurrency ecosystem, how it operates and its peculiarities;

additionally, I was immersed and active in the community as it underwent a very transformative time: 2017-2018 marked an era that urged us to define cryptocurrency as an asset class, respect the technologies that underpin it, and re-imagine how it fits into our lives, society and financial system.

Most importantly, I have extensively documented my journey with cryptocurrency from the beginning. I have a record of all challenges, questions, curiosities, concerns, opinions, failures, and successes. As cryptocurrency ventured towards the mainstream in 2017, I became excited as friends and acquaintances sought my help and advice to orient on the new technology and marketplace. Considering how far removed I was from beginners knowledge, I was glad to have documentation I could refer back to as I helped them avoid mistakes and streamline their learning.

Before long, I began consolidating all of my resources into one handbook that included everything I deemed necessary for mastering cryptocurrency markets and blockchain applications. As cryptocurrency and blockchain transformed throughout 2017 and 2018, I constantly renovated and supplemented the content to ultimately form the book you're reading today.

In this book, we will cover topics ranging from "the origination and purpose of cryptocurrency" to "navigating exchanges and evaluating investment opportunities."

If you follow this resource guide, I promise you will be awarded a solid foundation for informed investing and a springboard for further research.

AND, I promise that you won't go crazy sifting through the black hole of disjointed Internet articles and blog posts.

Cryptonomics is a one-stop-shop for beginners, curious minds, and even previous investors wishing to round out spotty knowledge of a complex and ever-changing topic.

Whichever group you identify with, I'm glad you are taking the time to educate yourself on a technology, community, and paradigm shift that will undoubtedly go down in the history books.

SECTION 1
A Comprehensive Introduction to Blockchain and Cryptocurrency

CHAPTER 1
A Technical Introduction to Blockchain

> *"Every fact of science was once damned. Every invention was considered impossible. Every discovery was a nervous shock to some orthodoxy."*
>
> — *Robert Anton Wilson*

THE TECHNOLOGY KNOWN as blockchain, and its relationship to cryptocurrency, can be a difficult topic to grasp.

Unfortunately, well-explained, accurate information regarding the function of this technology is lacking.

As the number of applications being built with blockchain continues to increase, it is more important than ever to gain a technical understanding of its design. By understanding how blockchain works "under the hood" you will be able to more effectively assess the ever-growing number of use cases and better understand the cryptocurrency industry as a whole.

This chapter promises to deliver this foundation in an

easy-to-understand fashion, while keeping the integrity and entirety of the concepts intact.

Blockchain, Cryptocurrencies, and Tokens

Blockchain is a digital ledger that records transactions securely, autonomously and without the need for any centralized third party. By using a variety of advanced security protocols (discussed later in the chapter), blockchain technology enables more efficient transfer and storage of transactions and the information contained inside them.

So how is blockchain related to cryptocurrency?

Blockchain is the underlying technology that enabled the creation of the first *cryptocurrency*, Bitcoin. When Bitcoin was founded, the engine that powered Bitcoin's entirely digital, pseudonymous, peer-to-peer payment system became known as the blockchain. Because Bitcoin was the first successful cryptocurrency to operate using a blockchain, it has persisted as the most widely used and referenced cryptocurrency, and is therefore the most important to understand. For this reason, we will use Bitcoin as the standard for understanding blockchain technology and cryptocurrencies.

The novelty of the Bitcoin payment network is that anyone can participate pseudonymously and all transactions are executed autonomously and without the need of a central bank or other intermediary. The founder of Bitcoin, Satoshi Nakamoto, believed a digital cash system free from any reliance on a central authority would solve several problems inherent to our current monetary system. Bitcoin, through its use of blockchain, proposed a new and potentially more efficient method of executing financial transactions.

Shortly following Bitcoin's creation, technologists began researching the potential for blockchain technology beyond digital payments, and soon realized its features could be utilized to better govern the transfer of information inside virtually any transaction.

Thus, entrepreneurs began using blockchain technology to create their own unique applications. Some of these blockchain applications were for other digital currencies that compete with Bitcoin, but many were created for entirely non-currency uses (such as identity verification or real estate transfer). In the next chapter, we'll delve deeper into the various applications being built with blockchain technology.

When a company is formed to begin developing a new blockchain application, it will often raise funds from investors for startup and development costs via an Initial Coin Offering (ICO), which operates similar to a crowdsale and will be covered in later chapters. Investors interested in the application being developed can purchase the cryptocurrency *token* of that company at a predetermined exchange rate, in hopes that the application will be successful and widely adopted, thus causing an appreciation in value of the underlying asset (the token). After the ICO is completed, investors can buy and sell the token on any exchange where it is listed.

Transactions on the Blockchain

Blockchain, at its core, is a technology that allows users in any network to exchange information with significantly less friction.

The stated purpose of blockchain may sound irrelevant and vague on the surface, but exchanging information quickly is at the center of modern daily life.

So how is blockchain better? After all, we already have systems and technology that allow us to govern the transfer of information, right?

This is true, but current systems require the help of a middleman or central authority to complete. For example:

- Buying groceries at the supermarket requires a credit card processing company and bank to intervene to move funds from your account to the grocery store and settle the payment

- Buying a house requires placing money in the hands of an escrow agent and hiring a realtor to close the sale

- Storage and retrieval of our own medical records requires the help of doctors offices, hospitals, or other actors in the health network

- Our most secretive and personally identifying information is owned and held in databases by government agencies or other organizations

What if these transactions could be completed without the use of a middleman?

Historically, we have enlisted the help of these trusted third parties because of the logistical convenience and added security of employing a central authority when dealing with the sensitive information contained in the above transactions. However, sometimes these central authorities prove expensive, ineffective, and untrustworthy.

What if we could enjoy the same convenience and security without having to trust that middleman or deal with their drawbacks?

Blockchain was created to be a technological "middleman" that acts as a fully transparent and auditable ledger and executes transactions based on a set of tamper-proof, codified rules.

Now, with a conceptual understanding of blockchain, the most effective way to understand its operating logic and potential use cases is by walking through a transaction from start to finish. Since Bitcoin boasts the original blockchain, which is used as a foundation for all successive cryptocurrencies, the following section focuses on a transaction in the Bitcoin network.

The Bitcoin Blockchain— An Example Transaction

Acquiring Currency and Creating a Wallet

Let's say Jim has digital currency he wants to send to Jane. First, he must create a digital wallet that allows him to store and send cryptocurrency. A wallet has 2 main components: a public key and private key. The public key, also known as a wallet address, identifies Jim and acts as a "location" where he can be sent Bitcoin (BTC); the private key is similar to a password in that it allows Jim to log in to his account and prove ownership of its contents. Anyone who becomes privy to this private key can access Jim's funds and produce falsified transactions. As the name suggests, Jim will want to keep his private key secret **AT ALL TIMES.**

Instead of the user creating public and private keys, like a username and password, they are generated randomly by the blockchain and appear as follows:

Public Key: 0x7FGJKD9ES0R

Private Key: FKDSW8450235KFJT

The pseudonymous public key allows Jim to participate in the Bitcoin network without using specific identifying information. Therefore, Jim can send/receive Bitcoin to and from anyone else in the network with only his public key string being displayed.

Creating a Transaction

To initiate a transaction, Jim will ask for Jane's public key and navigate to his wallet. After ensuring he has 1 BTC to send and confirming the wallet address provided by Jane is valid, Jim can confirm his transaction and send it to the blockchain for execution. Jim's transaction can be verbally described as "Jim sends Jane 1 BTC," but identities are not used in cryptocurrency, so this transaction will actually appear as:

1 BTC sent from Wallet address 0x7FGJKD9ES0R (Jim's wallet)-> *Wallet address 0xJDISJE8790J* (Jane's wallet)

Authorizing: First Hash and Signing the Transaction

Once Jim confirms his transaction, the blockchain starts to work and the transaction contents are sent through a "cryptographic hash function." A cryptographic hash function is a one-way encryption process that takes any given input and turns it into a random alphanumeric string. For example:

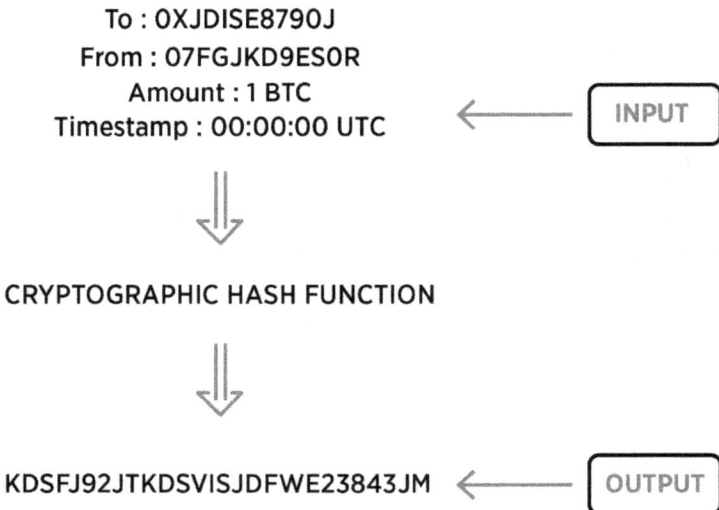

> To : 0XJDISE8790J
> From : 07FGJKD9ES0R
> Amount : 1 BTC
> Timestamp : 00:00:00 UTC ⟵——— INPUT

⇓

CRYPTOGRAPHIC HASH FUNCTION

⇓

KDSFJ92JTKDSVISJDFWE23843JM ⟵——— OUTPUT

The output "KDSFJ92JTKDSVISJDFWE23843JM" is known as the "hash value" and signifies the 1 BTC being sent from Jim's address to Jane's address at the specified time.

Since hashing is one-way, you cannot enter "KDSFJ92JTKDSVISJDFWE23843JM" (the output) and expect it to return "1 BTC sent from Wallet address 0x7FGJKD9ES0R -> Wallet address 0xJDISJE8790J" (the input). In this way, hashing is different from encryption, which is a two-way function.

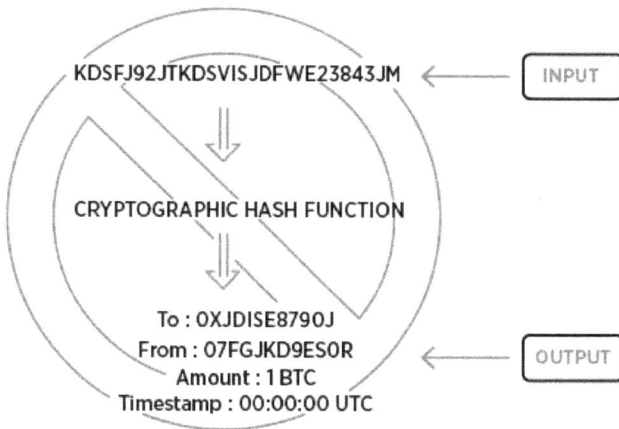

KDSFJ92JTKDSVISJDFWE23843JM ← INPUT

⇓

CRYPTOGRAPHIC HASH FUNCTION

⇓

To : 0XJDISE8790J
From : 07FGJKD9ES0R
Amount : 1 BTC
Timestamp : 00:00:00 UTC ← OUTPUT

This is similar to putting fruit into a blender and turning it into a smoothie; the process is irreversible.

With that said, it is also nearly impossible to guess the input that would return "KDSFJ92JTKDSVISJDFWE23843JM" as an output. The number of possible combinations is astronomical and there is no rhyme or reason to how a change to a given input may affect the corresponding output.

As you may expect, changing any part of the transaction will drastically change the output. Changing the transaction amount from 1 BTC to 0.99999 BTC will produce a new output that could look as different as "IOEVRTAHIOV8."

Once the transaction is hashed, Jim uses his private key to encrypt it. This is how we know that the owner of the account is the one altering its balance.

The outcome of transaction hash + private key is called a digital signature.

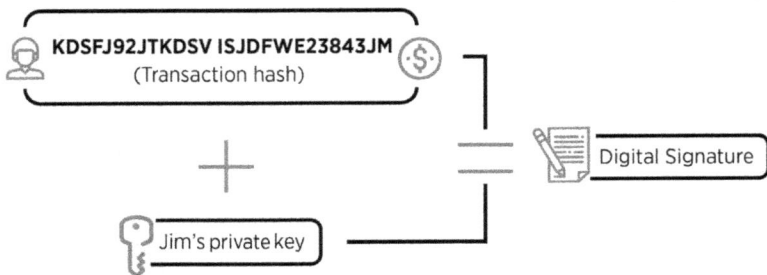

This digital signature is Jim's way of approving the transaction and binding himself to its contents.

When this transaction is broadcast to the network for confirmation (the next step), Jim can always verify that he owns this transaction by providing the digital signature instead of his private key. This is important, because digital signatures are innately unique to every transaction, since they are created by the combination of transaction contents *and* a private key. Since contents of a transaction will always be unique (no two timestamps are identical), digital signatures are also unique and never able to be reused.

This allows ownership of a transaction to be easily proven without divulging the private key that allowed its creation.

It is also important to note that it is not possible for someone, say Rick, to simply copy Jim's public key and create the following transaction:

Send 1 BTC from Wallet address 0x7FGJKD9ES0R (Jim's wallet) -> Wallet address 0x94DJVKDLDSKFJ (Rick's wallet).

After all, public keys are *public* knowledge and Rick could easily steal money from Jim's account this way.

However, when Rick attempts to "sign" the transaction with *his* private key, the blockchain will know it does not correspond with the originating wallet address/public key (since it is Jim's) and will not allow the transaction to happen. This prohibits users from authorizing transactions from accounts they do not own.

Luckily, this also means that Rick does not have to worry about another user initiating fraudulent transactions from his wallet address, either. This is the importance of the interrelatedness between a digital signature, private key, and public key.

The way these relationships work are central to the promise of blockchain.

These first 3 steps included the creation and verification of a transaction, but what if someone attempts to alter the contents of the transaction once Jim signs it? This would be similar, in concept, to Jim signing a blank check and subsequently allowing someone to fill in the "to" and "amount" line.

Fortunately, the blockchain has us covered as we're in transit, too.

If someone attempts to change the transaction contents once it is signed, the input will be altered and, therefore, a different output will be generated when it goes through the cryptographic hash function. In turn, the combination of that hash value and Jim's private key will return a different digital signature. That signature is checked against the real signature that we can verify came from Jim. If this happens, the real signature will be confirmed as legitimate and the fraudulent transaction attempt will be discarded.

Adding Transactions to the Block

Now that we know our transaction is secure, we can add it to a "block," which acts as a digital ledger containing hundreds or thousands of other transactions.

Hashing Transactions Together

Earlier in the chapter, it was stated that Jim's transaction contains 4 pieces of information.

1. To
2. From
3. Amount
4. Timestamp

However, each transaction actually holds a fifth piece of information: the hash of the previous transaction. This is how multiple transactions are linked together and chronology is established. Each transaction is added to the block in chronological order and uses the hash of the previous block as another piece of information that is mixed up to create the hash of the current block. Visually, multiple transactions hashed together looks as follows:

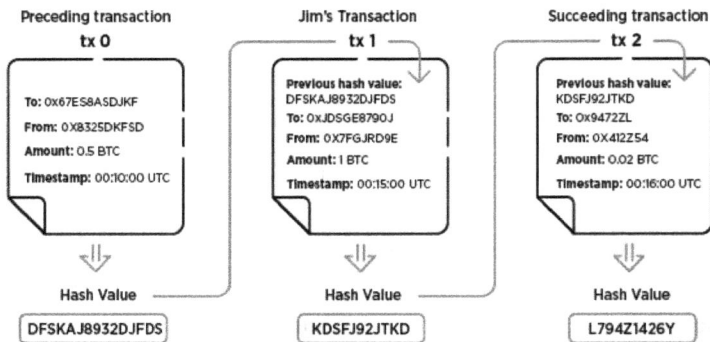

Preceding transaction	Jim's Transaction	Succeeding transaction
tx 0	**tx 1**	**tx 2**
To: 0x67ES8ASDJKF From: 0X832SDKFSD Amount: 0.5 BTC Timestamp: 00:10:00 UTC	Previous hash value: DFSKAJ8932DJFDS To: 0xJDSGE8790J From: 0X7FGJRD9E Amount: 1 BTC Timestamp: 00:15:00 UTC	Previous hash value: KDSFJ92JTKD To: 0x9472ZL From: 0X.412Z54 Amount: 0.02 BTC Timestamp: 00:16:00 UTC
Hash Value	Hash Value	Hash Value
DFSKAJ8932DJFDS	KDSFJ92JTKD	L794Z1426Y

Therefore, the hash value of Jim's transaction: (KDSFJ92JTKDSVISJDFWE23843JM) is used as a critical input to the next transaction.

Building and Hashing the Block

The hash of the most recent transaction on the block is called the "Merkle Root," and is one single hash value that represents all preceding transactions on the block. This Merkle root is included as one of five pieces of information on each block. All five pieces of information include:

1. **Merkle Root**: The one value that summarizes every transaction, and its internal data, on the block. This is the true meat of the block and allows us to look to one number for instantaneous verification of its contents.

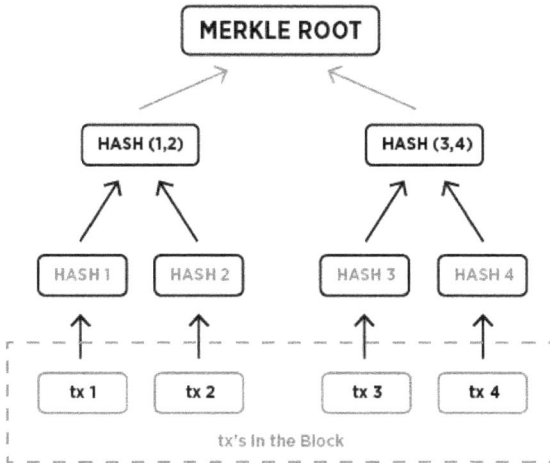

2. **Block Number**: This identifier starts at "the genesis block" (either 0 or 1) and counts up. At the time of writing, the current Bitcoin block is #509024.

3. **Timestamp:** In a few minutes, this block will join other blocks and become a part of the ever-growing blockchain.

This timestamp reinforces the chronology of blocks being verified.

4. **Hash of Previous Block:** Just as a transaction includes the hash of the previous transaction, a block includes the hash of the previous block. This links the blocks (and their contents) together and allows the ledger to record transactions in chronological order.

Remember, a block is simply a ledger that holds hundreds or thousands of transactions; so, in many ways, we are repeating a process similar to creating the transactions that comprise the block.

However, there is one main difference: in transaction hashing, five known inputs are used to generate a random, unknown output.

In block hashing, four inputs and a hash value (the ultimate output) are known, and the fifth input must be calculated to make the equation work.

Solving for this fifth piece, the "nonce", solidifies the block and officially adds it to the blockchain.

5. **Nonce:** A nonce is a random number that, when combined with the other 4 pieces of data on a block, returns the known hash value. Implementation of the nonce is central to the "Proof-of-Work" system that popularized blockchain and continues to differentiate it from other types of encryption.

With the nonce, the output is given, the block data is known, and we must find the value that completes the equation.

To find this nonce, individuals called "miners" use extremely high-powered computer systems (often referred to as mining rigs) to quickly increment millions of numbers until the correct value is found.

Miners invest in these computers and compete in the network

to find the correct nonce that allows blocks to be verified. The first miner to find the correct nonce then broadcasts it to other miners. At this point, all other miners can easily append that nonce to the block contents and quickly verify that it generates the predetermined block hash value.

If consensus is reached, the block is added to the blockchain. The miner (also referred to as a node) that identified the correct nonce is rewarded with a small amount of Bitcoin, thus the incentive to invest in mining rigs and the energy expended to run them. The infographic below provides a visual representation of hashing and linking blocks:

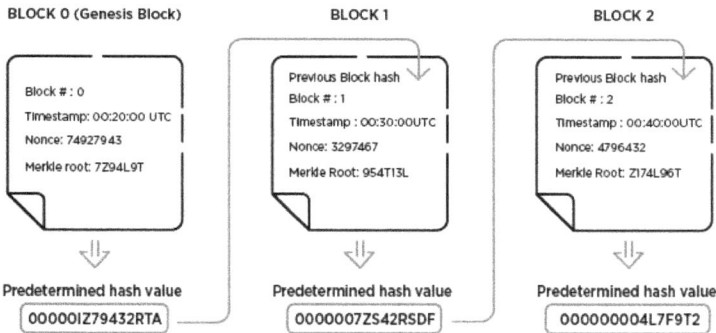

BLOCK 0 (Genesis Block)	BLOCK 1	BLOCK 2
Block # : 0	Previous Block hash	Previous Block hash
Timestamp: 00:20:00 UTC	Block # : 1	Block # : 2
Nonce: 74927943	Timestamp : 00:30:00UTC	Timestamp : 00:40:00UTC
Merkle root: 7Z94L9T	Nonce: 3297467	Nonce: 4796432
	Merkle Root: 954T13L	Merkle Root: Z174L96T
Predetermined hash value	Predetermined hash value	Predetermined hash value
00000IZ79432RTA	0000007ZS42RSDF	000000004L7F9T2

But what if an unethical miner attempted to change the contents of previously confirmed blocks to benefit themself? The creators of Bitcoin (and thus, the first iteration of the blockchain) wanted to ensure this was not easily done and any change to previous blocks required consensus of the whole network.

Therefore, the difficulty associated with finding the correct nonce is not random, and is frequently adjusted so that each block takes approximately 10 minutes to verify. Therefore, if someone changes the contents of a past block, it will cause a chain reaction and change the contents of _**all**_ successive blocks.

For example, if a miner attempts to change information on a

block that is 9 back from the one currently being mined, it would take approximately 90 minutes (9 blocks x 10 minutes each) to recalculate the nonce of all blocks that follow. This recalculation would have to be completed before the current block is verified, since the hash of previous transactions and blocks determine the contents of the current block. Since the current block will only take approximately 10 minutes to verify, the preceding blocks will never be successfully re-hashed in time by the hacker.

Once the nonce of the current block is found, it will become clear that a hacker is attempting to alter historical transactions and that hacker will be overruled, and eventually, booted from the network. Blocks that are verified with the incorrect nonce become "orphan blocks," as they are abandoned and the block-chain continues to grow with the correct nonce and succeeding block. We always accept the longest chain/main chain as correct.

Earlier, I stated, "It is nearly impossible to guess what input will generate a given output." On the surface, some may perceive the Proof-of-Work System and use of a nonce as an obvious contradiction of this assertion.

However, a nonce requires miners to obtain a value known to be numerical, which can be guessed and checked very quickly with a robust mining rig. In contrast, the multitude of inputs to a cryptographic hash function—and the uncertain nature of how modifying those inputs slightly may change the resulting hash value—allows for an astronomically greater number of possibilities that must be tested. The amount of computing power needed to iterate such a large quantity of possible values is currently unattainable.

Transactions on the Blockchain: A Review

This is the process of how a transaction is created and executed on the Bitcoin blockchain.

Let's recap:

1. Jim generates both a public and private key, which together constitute a unique wallet.

2. Jim creates a transaction. This transaction will contain data such as his wallet address (public key), receiver's address, amount, timestamp, and hash of previous transaction. This information is hashed and signed by Jim's private key. No one can sign this transaction without his private key, and, for this reason, Jim should always keep it secret.

3. By signing this transaction, Jim verifies its contents and binds himself to them.

4. Jim's transaction is added to the block and linked with other transactions in chronological order. There are thousands of transactions on each block and all the blocks are linked together to create a blockchain.

5. All transactions on a block can be represented by a single Merkle Root value. This root, along with the block number, previous block hash, and timestamp are the main pieces of information on a block.

6. Via the Proof-of-Work system, we are provided a predetermined hash of this block and increment a nonce (a random number that is only used once) until the correct hash value is found. This may take a million (or more) attempts.

7. This nonce, once found by a miner, is broadcast to all other miners. If miners reach consensus on the nonce, the block is verified and added to the blockchain.

8. It is very difficult to change previous transactions. In order to do so, the network must reach 51% consensus and re-calculate the hashes of each block that followed the one being modified. The older a block, the harder it becomes to hack or change.

Attributes of the Bitcoin Blockchain

Incorruptibility/Immutability

The blockchain is considered incorruptible because of the hashing process + Proof-of-Work system (using a nonce to reach consensus and solidify block contents), as well as other supporting mechanisms that guide the network. Ultimately, the blockchain is considered immutable because it is virtually impossible to change previous records once they are confirmed, since the continuously growing record of transactions becomes increasingly tamper-proof as the chain grows. In this way, the blockchain is inherently resistant to modification and unwanted tampering.

Distributed and Decentralized (most of the time)

All nodes in the network store and manage a copy of the entire blockchain. Each time the blockchain is changed, the ledger is updated across all nodes. For this reason, the blockchain is distributed between nodes and no single point of failure exists.

Most blockchains are also decentralized, meaning each user within the network has equal rights (access, permissions, and ability to verify transactions). However, some blockchains are designed to allow the end user to pick and choose which users have certain rights, thus centralizing its authority.

In the next chapter, we will discuss private blockchains, which often take on a centralized structure by designating an administrator to be in control. The difference between centralized, decentralized, and distributed ledgers is visually represented in the following figure:

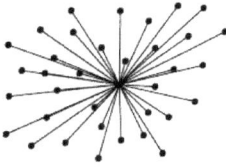

centralised   decentralised   distributed

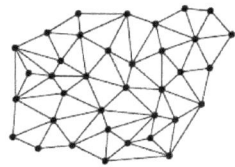

Many would argue that decentralization is the hallmark feature of blockchain. There is no central authority (a bank, business, trustee, or other organization) needed to verify and execute transactions. Therefore, this central authority cannot monopolize control or otherwise abuse their power. Perhaps more importantly, there is no single point of failure that can be compromised in the network. If a hacker attempted to infiltrate the system, they would be required to change the ledger held on every node and overpower the entire network.

Given the amount of computing power needed to outwork the Proof-of-Work mechanism and nonce, this is a near impossibility.

Trustless

With blockchain, we need not trust a central authority to have unwavering integrity or be error proof. Blockchain executes transactions objectively and automatically, with built-in mechanisms for flagging suspicious activity.

Additionally, some countries may have the luxury of trusting central authorities, but others do not. Developing, undeveloped or corrupt countries may suffer from oppressive, dishonest, or manipulative governments and businesses. The decentralized, anonymous nature of cryptocurrency and blockchain can be especially useful and liberating in these environments. The founder of Ethereum, the third largest cryptocurrency at the time of writing, is especially passionate about this feature and speaks on the topic often.

Autonomous/Objective

The blockchain has no incentive or ability to defraud since it is an objective software program with codified rules. It executes transactions according to its immutable code and these can only be modified with consensus from the network.

Pseudonymous

The Bitcoin blockchain allows for concealment of a users identity under a public key, thus no personally identifying information is needed to participate in the network.

*Note: any person can create and oversee as many wallet addresses as they desire. The balance of any wallet is public knowledge, so having multiple wallets can be a preventative security measure for anyone who feels uncomfortable storing a large amount of assets in one place.

Transparent

Since transactions are pseudonymous, the entire ledger can be transparent and entirely auditable without compromising the identities of users.

The entire ledger of transactions, dating back to the creation of the network, is readily available for all to see.

Digital

There is no need to physically print, store, or transfer money. The network and currency are digital in nature. This makes them more cost effective, scalable, and traceable. One critique of paper money is that it proves too prehistoric for modern society and does not take the peculiarities of E-commerce into consideration.

Low Cost/Free

There is no need to pay for expensive intermediaries to execute transactions.

This feature is passionately debated, since Bitcoin, and many other cryptocurrencies, require users pay a small transaction fee to the miners tasked with verifying each block. As Bitcoin has grown, many believe these transaction fees have become too expensive, and now violate a core tenant of its own original value proposition.

Deflationary

As we will explore in Chapter 3, Bitcoin was created in direct response to what Nakomoto perceived to be a failed monetary system. Therefore, the Bitcoin network was manufactured as a rebuttal to the interventionism and inflation inherent to our global economy.

While fiat[1] currency supplies are uncapped and can be throttled by the governments that own them, Nakomoto capped the Bitcoin supply at 21 million and built the Bitcoin blockchain to only allow for the creation of new coins when a Block of transactions is mined and the miner is rewarded. This is how the Bitcoin tokens are released into the economy for citizens' use.

By restricting the currency supply to a known and finite number, Nakomoto attempted to guard against the possibility of each unit gradually losing purchasing power due to ever-increasing supply (inflation).

Also, by restricting the release of coins into the economy, Nakomoto is reinforcing his assertion about the need for more organic (and democratic) management of our money supply.

---

1 Currency, such as the United States Dollar, that is declared by the government to be legal tender

As blockchain technology has matured and been matched with commercial needs, this feature has taken the back seat to most others in this list. However, the underlying philosophy has remained salient and most successive blockchains set a supply cap for similar reasons.

Blockchain:
Bringing it all Together

With these features in mind, we can identify blockchain as a protocol for business transactions.

Bitcoin, and the underlying technology it spawned, proposed a new way of exchanging value by taking trust out of the equation. The blockchain acts as a distributed ledger that efficiently records transactions between parties and relies on a unique mix of transparency, autonomy, and cryptographic trust to accomplish its task.

As previously stated, many features of the Bitcoin blockchain are interdependent and support each other. For example, concepts such as anonymity and transparency are typically viewed as contradictory, but in this situation they work together and are both imperative to the proper functioning of the technology.

Further, decentralization enables anonymity and transparency, but decentralization only works if the network is able to ensure that each node is unable (or discouraged) to corrupt the network. Thus, a Proof-of-Work system was implemented.

The true marvel of blockchain is in the unique way these various mechanisms are used in concert.

As you will see in Chapter 3, blockchain is effectively a creative combination of other technologies and systems that were pieced together by Nakamoto in a revolutionary way. He borrowed, renovated, and included concepts dating back to the 1960s in order to create the solution we now know as blockchain.

As Bitcoin gained coverage and adoption, other developers began thinking about how blockchain could be used to secure all types of transactions, not just ones that involve transferring digital money.

Naturally, this lead to several major improvements and updates to the original blockchain.

For example, blockchain developers have since created other widely accepted consensus mechanisms such as Proof-of-Stake and Proof-of-Authority, which each boast their own unique way of securing and executing transactions based on the foundational tenets of the Bitcoin blockchain. Neither of these iterations would have been possible without the original Proof-of-Work mechanism, and none is universally recognized as superior to the other; rather, blockchain developers choose the appropriate features based on the application and goal of the project.

Final Words

Today, companies are developing blockchains for applications ranging from identity verification to smart contract execution. Many second movers proposed payment systems to compete with Bitcoin based on improvements made to its blockchain; however, it is very important to note that the cryptocurrencies with a true currency application are in the minority.

Today, there are over 2,000 cryptocurrency tokens available for purchase on cryptocurrency exchanges around the world, and each token is connected to a company creating their own unique blockchain application.

In the next chapter, we will completely examine these application categories and their potential.

CHAPTER 2
The Taxonomy of Cryptocurrency and Blockchain Applications

IN THIS CHAPTER, we will move beyond the Bitcoin blockchain to fully understand the context of what it has created. While the first chapter focused on the technology that underpins cryptocurrency tokens, this chapter will focus solely on the user-facing applications being created for real-world use cases.

Today, the technology fueling Bitcoin is being used to innovate for many of the world's most prominent industries. This chapter will more fully explain the applications of blockchain, examine the promise of each use case, and drive home the importance of understanding the connectedness between the underlying technology and investible cryptocurrencies that represent them.

First, it is important to understand the two main classifications of blockchains.

Private vs. Public Blockchains

The main distinction between public and private blockchains is related to who is permissioned to participate, execute the consensus mechanisms, and store a copy of the shared ledger.

Public blockchains are networks able to be joined by anyone. In fact, many public blockchains offer incentives, such as monetary rewards for mining blocks and helping maintain the ledger. The Bitcoin blockchain described in Chapter 1 is a quintessential example of a public blockchain: anyone can join the network, view/use the source code, and become a miner by solving the cryptographic puzzles to help maintain the network. These blockchains, and their records, are entirely open and auditable. While public blockchains serve an important purpose, some fear they are currently impractical for certain enterprise applications.

Private blockchains often have similar core infrastructure to their public counterparts, but require an invitation to join and participate from the network administrator. Typically, there are rules that restrict who can participate, view certain transactions, execute the consensus protocol, or maintain a copy of the ledger.

Private blockchains are faster and allow the user to exclude certain nodes from the network; however, they are less secure and powerful than public blockchains. In addition, many critique private blockchains for violating a core tenant of blockchain: balanced decentralization. By choosing who can participate and validate the network, a private blockchain is not truly balanced and maintained across all nodes, thus centralizing the power to a single point of failure. Nevertheless, the current development and testing of private blockchains is an important step towards bringing this technology to industry and will deliver drastic improvements in back office functions, data visibility and data integrity.

The Taxonomy of Cryptocurrencies: Distinguishing Currencies, Base Protocols, and User-Facing Applications

In this section, we will further explore the taxonomy of crypto-currencies by grouping them into 3 major categories: *digital currencies, base protocols,* and *user-facing applications.*

1. An example of a true ***digital currency*** is found in Bitcoin. Its intended application is as a currency, which individuals can use as a store and exchange of value, just like fiat money.

2. Any cryptocurrency categorized as a ***base protocol*** is building a blockchain from scratch and therefore not dependent on other blockchains to operate.

   Most new blockchain applications being created are not built from scratch, but rather borrow pieces of other trusted and proven chains to build their core infrastructure. With these base protocols as a solid foundation, teams developing applications in their infancy can build out the distinguishable front-end features of their application with less time and resources devoted to back-end construction.

   By standardizing much of the common infrastructure between blockchain applications, and offering it for drag-and-drop use, base protocols have enabled the rapid iteration and testing of blockchain applications.

   A visual representation of a base protocol's utility is shown in the following figure:

The Web

Blockchain

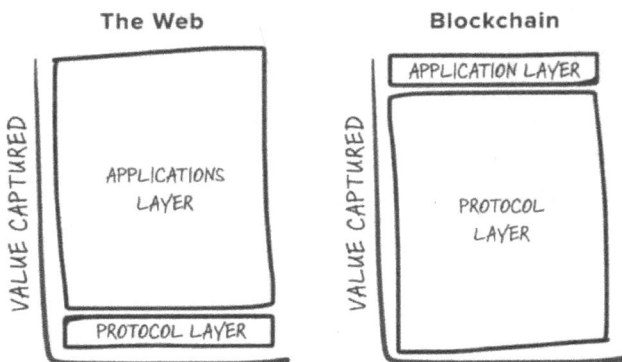

The base protocol layer of a blockchain is considered "fat," thus minimizing the amount of re-work performed by the architect of any new application. By comparison, the relatively thin protocol layer of the Internet demands more structural legwork from the developer of any new web application.

Due to the fidelity required of base protocols, development is time, capital, and human resource intensive. For this reason, the development of new base protocol applications is uncommon.

Currently, Ethereum and Neo are two of the most well known and commonly used base protocols and are universally trusted due to their track record, robustness of code, and number of applications supported. New projects or ICOs will often broadcast their underlying protocol layer to assure potential investors and adopters of the safety and effectiveness of their blockchain.

While the various base protocols serve a similar purpose, they are not all created equal. The architect of any new project, or decentralized application (dApp), will consider the specific advantages and drawbacks of base protocols to ensure technical specifications are sufficient for the goals of the project.

Since base protocols act as the building blocks for the entire ecosystem, they are typically recognized with their own unique classification. However, the use case category of "base protocol" could technically also be considered a user-facing application.

3. The *user-facing* application category is a catchall for all other dApps being built on a blockchain for a utilitarian purpose. The goal of a company building a user-facing application is typically to achieve adoption of their product commercially, in the same way any other innovative software would be adopted and utilized by an individual, company, government, or other entity. A sampling of popular application categories is explored in the next section.

Application Use Cases + Examples

Smart Contracts

Smart contracts are intentionally listed first because of their foundational importance to all other blockchain applications.

As a reminder, the core purpose of blockchain is to facilitate frictionless transactions between permitted parties in a trustless and decentralized manner, thus saving time and resources. Likewise, the term "smart contract" encompasses any circumstance in which a computer protocol oversees and enforces the faithful execution of a two-sided agreement.

This is achieved by using an "if this, then that" function.

For example, the buyer of a house may agree to purchase it for $200,000, pending successful completion of agreed-upon work by the seller.

Usually this requires a third party or escrow service to ensure that both sides hold up their end of the deal before executing the transaction. Additionally, it may require lawyers, brokers, or agents (all middlemen that cost money). A smart contract operates by allowing both parties to agree upon and input conditions to a digital middleman, which then executes the transaction when all conditions become satisfied. Once the conditions are met, the smart contract will automatically handle all closing and settlement activities.

An entirely different, but equally promising use case is voting: many have proposed that we can revolutionize our nation's voting system (and raise embarrassingly low turnout in the process) by leveraging smart contracts to streamline the polling process, reduce associated paperwork and combat voter fraud with block-chain-based identification systems. Currently, blockchain-based voting applications are being piloted on a smaller scale by local government or within organizations that use proxy voting.

At their core, smart contracts allow us to remove the middleman from a transaction. Conceptually this isn't much different than, say, an order processing system on a website that acts as an intake for orders and automatically sends them to the manufacturing floor for assembly. This order processing system replaced what was once a less efficient process, in which a human was required to take orders from customers and manually schedule them for production.

In many ways, blockchain technology offers an improvement to many of the systems that previously played a role in automating or streamlining a cumbersome process.

It is important to note that the idea of a smart contract—digitizing and decentralizing a relationship between two parties, thus taking trust out of the system and streamlining a trans-action—is foundational for almost all other applications.

In the following sections, you will notice that many successive applications operate with this same underlying philosophy, just with different specific audiences or uses in mind.

Asset Digitization/Supply Chain Visibility

Real-time asset digitization is relevant to companies who ship valuable goods or desire better visibility to their inventory.

Asset digitization improves the tracking of products while in transit by allowing for automatic and immutable transfer of shipping details to all parties in the supply chain, at each step of the journey.

This shipment information may include record of sourcing location, stops taken along the journey, hands changed, and environmental conditions while in route.

Obtaining access to these details could be the difference in a company ensuring the integrity of their product, capturing supply chain efficiencies, and reducing bottlenecks.

Consider the following examples:

1. *Supply Chain Visibility:* A luxury retailer is able to track the chain of custody as their designer handbags make their way from manufacture to point of sale. Retailers can ensure the product received has not been tampered with and is authentic, since it is assigned an unchangeable unique identifier and can be tracked throughout the entire supply chain.

   Blockchain Examples: VeChain, Waltonchain

2. *Sourcing:* Tracking a diamond from mine extraction to purchase. All parties involved (including the consumer) achieve peace of mind in knowing that the stone is authentic and ethically sourced.

   Blockchain Examples: Everledger, TrustChain

3. *Medicine:* Tracking the environmental conditions of certain medicines while in transit to the end patient. Some medicines or materials must be kept at a certain temperatures, humidity levels, or treated with other special care while in transit.

Automatically tracking and making this data available to necessary parties helps ensure transparent record keeping, compliance, safety and reliability.

Blockchain Examples: MediChain, Medicalchain

4. *Agriculture:* By automatically collecting data on soil conditions, environment, and pesticide use, agriculture companies can make better decisions on planting methods, improve the quantity and quality of production, and minimize the use of certain fertilizers and pesticides.

Blockchain Examples: AgUnity, Origin Trail, AgriDigital

In all of the above scenarios, an immutable, fully transparent, and auditable record of transactions helps an organization ensure the quality of products sourced as well as adherence to ethical sourcing practices. Likewise, this ledger could be easily audited for suspicious activity by a regulatory body.

In such a connected world, we increasingly rely on trust mechanisms to ensure the safety and authenticity of goods used daily as they travel across the globe. Augmenting (or entirely overhauling) outdated mechanisms with blockchain could provide a marked improvement to countless areas of our national and global supply chain.

Asset digitization applications are designed to protect the integrity of everything from life saving medicine and designer handbags to food, as they change dozens of hands, ride in ships, planes, and trucks, and are delivered to the end customer for eventual use.

Currencies

Currency applications, such as Bitcoin, were created for use as a store and transfer of value, similar to fiat currency. The project creators intended that the attached token be used for this explicit purpose.

While the coins that comprise this category have many similarities, most companies will build their use case around significantly modifying a core tenant of the original Bitcoin payment network concept.

For example, ZCash is widely recognized as the "privacy coin" and supports semi-transparent transactions, meaning only select transaction data is shared with other nodes in the network. Another example is found in Bitcoin Cash, which is a fork[2] of Bitcoin and proposes increasing the size of each block to improve transaction speed and decrease fees paid to miners.

While Bitcoin still controls most of the currency application market, several currency applications could have their own unique role in the digital economy moving forward.

Exchanges

Some, but not all, cryptocurrency exchanges are blockchain projects. The exchanges with purchasable cryptocurrencies supporting their network are blockchain applications. The blockchain project of the exchange will often include services such as decentralized, peer-to-peer (P2P) lending for traders looking to trade on margin, or underwriting services for upcoming ICOs.

Fintech

Fintech is a broad category that encompasses all finance-related blockchain projects. This includes applications such as peer-to-peer payments, cross-border payments, decentralized lending, and decentralized Crowdfunding.

---

2 When a core group of developers decides it would be best to change a central piece of a tokens operating code, thus creating a new and entirely independent token and blockchain. Forks are explained further in Chapter 4.

Examples of specific Fintech blockchain applications include:

- *International Payments and Remittance:* Historically, the ability to transfer money internationally has been synonymous with excessive processing times and fees; however, applications such as Stellar are building blockchains to facilitate frictionless, immediate, and affordable transfer of money across borders.

- *Clearing and Settlement:* Any active stock market investor knows about the T+3 rule (more recently T+2), which dictates that ownership of a security does not actually change hands and "settle" until 2 days following the date of the transaction. Given the blockchain's ability to automatically execute transactions, it could provide a platform for achieving clearance and settlement instantaneously in capital markets. Recently, investment banks and other financial institutions have expressed support of these applications via advocacy, investment, and adoption.

Database Management

Database management includes any application in which blockchain helps streamline the organization, storage, protection, and retrieval of data. In a similar way to smart contracts, this represents the "bread and butter" of blockchain's initial stated purpose.

Today, nearly all organizations house information in databases for safekeeping and seamless access by permissioned parties. The government, for example, manages the holy grail of sensitive data—personal information for millions of individuals. Primarily, they are concerned with the safe keeping of this data. However, they may also benefit from enhanced organization and retrieval systems, due to the significant administrative processing required for many government activities.

Example activities may include obtaining permits or assessing security clearance requests. The process of completing these tasks could be at least partially aided by a rules-based consensus protocol such as blockchain.

The government is only one of several entities to benefit from managing data with blockchain. This application could find a home in any company or organization that deals with large stores of data, especially if that data is sensitive in nature. Today, many companies are researching, or at least interested in, retrofitting blockchain to manage data generated by their operation.

In light of several serious data breaches throughout the past several years (Sony, Target, Equifax, Facebook), database management applications of blockchain are emerging as a potential solution to this undeniably pressing issue.

Identity Verification

In the Equifax hack, millions of identities were mined from a hack on the company's internal database. This means all information given to Equifax, down to the social security number, was compromised.

Given the increased frequency and severity of these hacks, the promise of this application is undeniable.

Blockchain-based identity verification systems allow for decentralized storage of data and give the user freedom in granting access to third parties as needed. This could include permissions ranging from age verification at a bar to providing a social security number for checking a credit score.

This data can be secured, managed, and warehoused better. Data managed on a blockchain, versus a centralized database, assumes the characteristic of sovereignty, making it readily available to permissioned parties without inviting the risks inherent to centralized storage.

This decentralized structure ensures the protection of sensitive information, since would-be hackers are not able to concentrate all resources on corrupting only one node in a data heist.

Fundamentally, companies building identity verification applications are attempting to solve a conundrum referenced in the first chapter—achieving transparency and privacy simultaneously.

Gaming/eSports

In recent years, the eSports arena has increased in complexity, as professionals win sponsorship deals and teams recruit and manage new talent to compete for prize money in tournaments. This application class may not seem as large in scope as others, but the eSports community maintains an affinity for emerging technology and has historically acted as a testing ground for new innovations. Blockchain companies operating within this space are creating solutions to streamline the logistics and administrative tasks of these (sometimes very involved) tournaments. This may include facilitating prize payments to teams, decentralization of sponsorship deals, and even allowing seamless in-game purchases.

Blockchain Applications: A 30,000-Foot View

The above list of applications contains a mere sampling of dApps being created today.

While use cases may vary based on the industry of interest, the value proposition for all dApps is the potential for better management of digital relationships by enabling the secure, efficient, and effective transfer of data using a trustless and decentralized system.

These applications have the potential to enhance the way any given industry tracks, retrieves, and manages data.

The Greater Cryptocurrency Landscape

Along with the ever-increasing number of decentralized applications being built, there are an ever-increasing number of entities playing a supporting role in the general development of the ecosystem.

These include:

- Exchanges that generate income by charging brokerage fees to investors who place trades on their platform. Coinbase, Poloniex, Bittrex, and Binance are popular examples of cryptocurrency exchanges

- News outlets and journalists focused on cryptocurrency and blockchain

- The emergence of widely recognized experts and thought leaders

- Research labs created to study cryptocurrency and block-chain via collaboration among corporations, higher education institutions, and governments

- Classes and training programs offered online or at traditional universities

While the development of the actual technology and applications is crucial, it is also important to recognize how these peripheral entities are attaching themselves to the emerging space, oftentimes acting as a bridge to connect an immature market with mainstream adoption. Especially in 2017 and 2018, there has been a noticeable convergence of both blockchain and cryptocurrency with established entities.

In Chapter 7, we will dive deeper into the progress being made in the development of applications such as those outlined above.

Final Words

Blockchain and its uses can seem esoteric.

This chapter endeavored to familiarize the reader with real-world, relatable examples of the commercial value contained inside the technology popularized by Bitcoin.

Today, this asset class has grown to incredible diversity and delivers value in more ways than simply in the form of digital currency. In fact, much of the current interest and investment is in non-currency, and even non-financial, applications.

Of course, not all applications are created equal; therefore, Chapter 6 is dedicated to learning a framework for evaluating the promise of various projects.

For now, in the next chapter, we will shift our focus to the history of cryptocurrency and blockchain through the lens of the industries they are poised to disrupt.

CHAPTER 3

The History and Evolution of Cryptocurrency Markets

> *"There is nothing new in the world except the history you do not know."*
>
> —*Harry S. Truman*

THIS CHAPTER WILL be broken into 2 parts:

First, we will discuss the brief history of blockchain as a technology.

Second, we will take a trip back in time to explore the more lengthy history of the financial system that led to the creation of Bitcoin and the entire cryptocurrency ecosystem.

The [Short] History of Blockchain

Although blockchain is a very new and innovative technology, its core operating principles contain remnants of science, information systems, and economics that date back decades.

As you read in Chapter 1, blockchain has roots in encryption, which is a branch of cryptography.

Hence the *crypto* in *crypto*currency.

Encryption is simply a way to disguise sensitive information under an unrecognizable code that can only be translated by using a key to decipher. The marriage between digital systems and encryption came shortly after the invention of computers. In fact, one of the main uses of early digital computers was to break ciphers (coded messages) sent during WWII. More recently, the creation of Bitcoin was inspired by several cryptographic systems discovered decades ago.

For example, public key cryptography was discovered in the 1970s and the Proof-of-Work consensus mechanism was proposed in an article published in 2002.

Today, many organizations are iterating Satoshi's original idea of blockchain to create their own specific applications.

Admittedly, the history of blockchain could be further explored to include various developments in fields such as cryptography and data structures, but that remains outside of this book's primary focus.

Instead, this book elects to favor the economic and financial history that conditioned the rise of Bitcoin and cryptocurrency markets.

The [Slightly Longer] History of Cryptocurrency

The inception of cryptocurrency follows a buildup of events over several decades, which ultimately led to the publication of Satoshi Nakamoto's white paper[3] in 2008.

---

3 An informative publication issued to educate readers or state an opinion on a topic. White papers are a common tool used in business to present research or case studies to an audience in an attempt to inform a decision.

While the Bitcoin payment network outlined in Nakamoto's white paper was impressive, he/she was not the first to propose a payment network governed by cryptographic trust. Others had previously attempted to create similar applications and failed. These failures were attributable to a few distinct areas:

1. The proposed systems relied on a bank or other central authority (lack of decentralization)

2. The proposed system ultimately relied on its creator to secure and execute transactions. When the creator was no longer able to keep up with the demand of the network, it crumbled (lack of decentralization and autonomy)

3. Mainstream adoption wasn't achieved due to a lack of perceived need for the innovation at the time (timing is very important)

So what made Nakamoto and Bitcoin different? The answer comes from unraveling a financial and economic history of more than 100 years.

The Gold Standard and Foreign Exchange Market

In 1900, the United States established the *gold standard*, which dictated each dollar was redeemable for a fixed amount of gold. This ensured the nation's dollar was backed by something tangible and universally accepted as valuable, while also limiting the government's ability to print money at will. This standard-ization also helped facilitate trade between countries, since gold could be used as a common measuring stick for value and was accepted globally.

However, the onset of the Great Depression forced many countries around the world to abandon the gold standard and allow for

---

In cryptocurrency, new businesses will often compose white papers, which include an outline of the problem being solved and background of the solution proposed by the team.

unfettered injections of paper currency into the economy as deemed necessary by the government. Many economists cite this dissolution of the gold standard as necessary at the time and a major contributor to weathering the Great Depression in the 1930s.

From the 1930s through the end of WWII, the United States operated on a quasi-gold system and was selective with who could exchange currency for gold.

After enduring such an economically taxing war and Recession, a return to the gold standard became less likely, as governments claimed a continued need to manage the flow of money into the economy.

Post WWII, international trade encouraged the mingling of economies as several macroeconomic changes led to the creation of organizations such as the World Bank and International Monetary Fund (IMF) to assist in regulating the global economy.

This time period and its effect on our current economy is a book in itself. For our purposes, it is important to note two things:

1. At this point in history, our currency no longer holds true intrinsic value. It has no more actual value than a piece of paper. Fiat money is accepted universally because it is deemed "legal tender" and backed by the government as the official currency of a certain nation-state.

2. Due to the increasingly global nature of commerce, the world economy required a new system for exchanging fiat currencies. Since all countries stopped underpinning their currency with gold, we lost our common measuring stick, thus inviting a need for the establishment of a market that maintains the exchange rates between currencies.

The *foreign exchange market* is an entirely digital marketplace for trading fiat currencies. The value of any currency is expressed relative to another currency. For example, at the time of writing, the Euro/United States Dollar exchange rate is 1.23/1. That is, investors need 1.23 USD to purchase 1 Euro. This means the euro is "stronger" than the dollar, since it takes more than 1 USD to obtain 1 Euro.

The foreign exchange market has four main uses:

1. It enables companies and organizations to engage in global commerce.

2. It allows individuals to extract value from their native currency when visiting foreign nations.

3. It acts as a platform for financial institutions and individual traders to speculate on the future price of currencies and make profits in the short term.

4. It is used by central banks and other large entities wishing to influence certain macroeconomic conditions by trading large amounts of currency in the open marketplace.

This "management" of currency price by various actors in the marketplace has led to the creation of a massive, and often volatile, foreign exchange market. In addition, the implication that any given nation's currency value may be manipulated by powerful market participants elicits skepticism from some regarding the fairness of the market and stability of their nation's money.

The Bubble (Early 2000s and Dot-com)

Many investors are aware that the rise and fall of the Dot-Com era (or, collectively, the Dot-Com bubble) of the early 2000s is frequently paralleled to cryptocurrency markets.

The Dot-Com era began shortly after the turn of the millennium, as Y2K and the development of the worldwide web urged humans to re-define their relationship with computers and the digital world. The Internet would soon become positioned to revolutionize computation, connectivity, commerce, and much more.

This time period marked the dawn of the Information Age, as companies and individuals rushed to capitalize on the creation and growing popularity of the Internet. Many entrepreneurs tried their hand at creating internet-based applications, and investors rushed to get in on rapidly growing dot-com businesses, creating mania in the capital markets.

Unfortunately, the market growth experienced in this era was perpetuated by hype from both ends. Many dot-com companies allocated incredible sums to advertising instead of building the actual application, and it seemed the market and investors automatically perceived any organization that slapped ".com" to the end of their name as more valuable. Likewise, investors were too easily convinced and didn't conduct sufficient research. Both hype and asset prices soared until they reached a point that could not be sustained.

Investors became skeptical due to the meteoric rise in asset valuations, but a lack of actual value generated to justify the increases. Infamous companies, such as Pets.com, were forced to declare bankruptcy due to blatantly unsustainable financial models.

Pets.com wasn't alone. Countless other companies who failed to generate meaningful traction were hiring hundreds of employees, spending tens of millions of dollars in advertising per year, and boasted market valuations that no logical financial analyst could defend. All of these companies would eventually take a hit and even some with potential could no longer sustain themselves due to the blowback from the market. However, some of the most influential companies in existence today survived this bubble burst and came back stronger.

This bubble acted as a rupture to weed out overvalued and underperforming companies, check the unchecked optimism, and move forward to embrace an exciting technology in a more sensible way.

Sound familiar?

It has long been debated whether cryptocurrency has experienced, or is currently experiencing, a bubble similar to Dot-Com. Frankly, it appears easy to draw similarities: the excessive valuations; the mysterious and digital nature; the mania surrounding anything claiming to be a cryptocurrency or have a blockchain; a palpable fear of missing out (FOMO); and the obnoxiously fast pace of the market.

The popularization of these parallels has provided enough of a reason for many investors to avoid the cryptocurrency markets altogether.

However, insights gleaned from the Dot-Com bubble help us understand the ebbs and flows of exciting new technologies such as blockchain.

More importantly, these insights remind investors to be judicious and practical when considering emerging markets.

Invariably, capital markets began recovering shortly after the Dot-com bust. However, just as markets appeared to race towards pre-bubble levels, the economy was dealt one of the most damaging blows since the Great Depression.

2008 Recession

The Great Recession initiated a global economic meltdown and exposed a precarious and misguided Wall Street to the public.

This recession can be linked, at least in part, to a failing bond market.

A bond is one of many assets available for purchase by investors and can be backed by a company, government, or other entity.

One class of bonds available is comprised of individual home mortgages. As a homeowner pays back their mortgage (plus interest) to the financial institution that provided the initial loan, the investor who owns the bond is paid back, in turn.

The 2008 Recession was partially or entirely due to a misrepresentation of the quality (and, therefore, risk associated with) these bonds. Banks made a habit of selling bonds backed by very low quality mortgages: some of the mortgages underlying these bonds were known to be "subprime" and contain a high default risk, due to historical delinquency of the borrower. After a while, the mortgages became due, the borrower couldn't pay, and the bonds stuffed with these poor quality mortgages collapsed.

Both homeowners and market participants lost their investments, and the banks that packaged these bad bonds were scrutinized for abuse of power and unethical practices. This added fuel to the fire of the housing crash and conditioned a negative perception of Wall Street as greedy and untrustworthy. Ultimately, criminal probes were conducted and several banks faced legal recourse.

This market crash would ultimately serve to rupture trust between consumers and the financial establishment.

After the dust settled and the nation witnessed the bankruptcy of seemingly untouchable firms such as Lehman Brothers, it became clear that this market crash marked the end of an era in the financial world.

The Creation of Bitcoin

Shortly after the Lehman Brothers bankruptcy announcement, in October 2008, a white paper emerged entitled *Bitcoin: A Peer-to-Peer Electronic Cash System*. At a time when our *trust* in the current system was lowest, Satoshi Nakamoto, the author of the white paper, offered a seemingly far-fetched solution that took *trust* out of the equation.

Nakamoto proposed an entirely decentralized and trustless financial system, in stark contrast to the powerful, centralized system that had freshly disappointed citizens across the globe. Bitcoin boasted features that guarded against problems such as government manipulation, currency devaluation, and centralization of ownership. With these features, Nakamoto carefully positioned Bitcoin as the antidote for a poisoned financial system.

Satoshi Nakamoto

Satoshi Nakamoto is the pseudonym used to reference the person or group of people who founded Bitcoin. A few individuals in the computer science and financial industries have claimed ownership of the pseudonym, but those claims have been largely debunked as false.

Due to the cross-disciplinary nature of Bitcoin, drawing from functions including finance, computer science, and statistics, experts believe Nakamoto to be either a highly capable group of individuals or a lone-wolf genius aided by substantial early help from supporters of the project.

Bitcoin: The Early Years-Present Day

In the early days of Bitcoin, Nakamoto used this pseudonym to collaborate with developers through Bitcoin.org and online forums. In early 2009, the first software for Bitcoin was released and the genesis block of the Bitcoin blockchain was mined.

Since Nakamoto released a robust first version of the Bitcoin software in January of 2009, it is believed that work on the project began as early as 2007.

Over the next few years, Bitcoin's adoption continued increasing while other cryptocurrencies were created in response. The first transaction (10,000 BTC for two pizzas) was successfully executed in 2010. Companies such as BitPay, a payment processor, and Mt. Gox, a well-known exchange, were founded, creating infrastructure around this new digital currency. A few daring businesses, small and large, began accepting Bitcoin and other cryptocurrencies for products and services.

Experts in finance and computer science began debating about Bitcoin and it's potential. Adoption and trading volume increased as the price of BTC rose into the hundreds of dollars, until one of the main exchanges at the time, Mt. Gox, announced it was experiencing significant technical issues.

In 2014, shortly following this announcement, Mt. Gox was hacked, lost hundreds of thousands of Bitcoin, and subsequently filed for bankruptcy. Many early Bitcoin investors likely had their BTC compromised in this hack, another malicious attack, or by faulty error, as early Bitcoin wallets were less user-friendly than today, and many were less secure or more prone to catastrophic user error.

Around this time, the FBI dismantled the infamous Silk Road black marketplace(s) and seized approximately 144,000 BTC in connection with illegal activity. For a period of time, this FBI seizure gave the agency ownership of the largest Bitcoin wallet, but the contents of that wallet have since been sold off in numerous government auctions. These two events (Mt. Gox and Silk Road) dealt a blow to Bitcoin's momentum and caused many to permanently associate it with nefarious activity and poor security oversight. To date, this label has proven difficult to shake, and many are unable to look past these events to consider the promise of the asset class and underlying technology.

Fortunately, as the marketplace has matured, service providers are becoming more proactive about protecting against scams and hacks with better application design, enhanced security protocols, and better user interfaces.

Blockchain and cryptocurrency began accelerating once again towards the beginning of 2016, as new altcoins (a designation given to all cryptocurrencies that have followed Bitcoin) emerged at a rapid pace.

During this time, the scope and frequency of cryptocurrency news coverage increased, revealing both positive and negative attributes of the technology and asset class. This journalistic interest coincided with, and possibly facilitated, the teeming curiosity amongst governments and enterprises.

In 2017, awareness skyrocketed as blockchain and cryptocurrency captured mainstream interest across the globe. This fueled the market mania experienced in 2017, as we saw the price explode through the thousands and surpass the elusive $10,000 per BTC mark. This heightened interest also forced the public to more thoroughly research and recognize blockchain as a potentially revolutionary technology; thus, many groundbreaking applications were brought to center stage.

2017 was undoubtedly a pivotal year for cryptocurrency and blockchain. Inside the community, early adopters and newcomers alike expressed excitement for the increase in activity and interest, both in the financial marketplace and on the application front. Likely due to this excitement, the iterations of Bitcoin and other ICOs exploded.

The industry experienced its fair share of growing pains, too. The fast pace and uncertainty in the marketplace demanded a high tolerance for volatility. Unfortunately, many novice investors jumped in too quickly without sufficient due diligence, and, as a result, fell victim to ICO scams, hacking, phishing, and the type of emotion-driven "investing" that was always bound to end badly.

Today, all blockchain projects and their connected tokens have followed unique trajectories.

2018 dealt the cryptocurrency markets one of the largest corrections to date, and while the market has exhibited promise, it continues to be caught in the middle of an uncertain regulatory landscape, persistent fraud, clever scams, and vicious security compromises. Blockchain adoption, on the other hand, has persisted at an impressive rate.

The challenges and progress inherent to this type of emerging marketplace are further explored in the final chapter.

Final Words

While the first half of this book endeavored to orient the reader on cryptocurrency as an asset class, how it came into existence, and who it may benefit, the next 3 chapters will shift focus to exploring the research methods and tools used when navigating the cryptocurrency ecosystem.

Finally, the last chapter will synthesize all previous information into an analysis of the current state and proposed future of blockchain and cryptocurrency.

SECTION 2

Navigating the Cryptocurrency Ecosystem

CHAPTER 4
Investing in Cryptocurrencies: Exchanges, ICOs, and Mining

In 2017, the cryptocurrency markets experienced unprecedented interest and massive volatility. As a result, we saw waves of new entrants eager to capitalize on the trend, yet a shortage of reliable resources made it extremely difficult to safely access this new market.

Before learning the prevailing methods for evaluating these tokens, investors must first become comfortable with navigating the ecosystem safely.

As is true of all other asset classes, the cryptocurrency ecosystem has its own unique rules, norms, and peculiarities. Not becoming properly oriented can lead to confusion and lack of productivity at best, and loss of your investment at worst. The following two chapters, especially, act as a streamlined resource guide for everything an investor needs to know about getting set up to invest in the cryptocurrency markets.

These chapters will cover topics ranging from buying and storing cryptocurrency to navigating research repositories. Most importantly, they will focus on performing these activities *safely*.

Similar to the format of the Chapter 1, these chapters will examine the cryptocurrency marketplace by walking through a transaction—in this case, the life cycle of purchasing and storing cryptocurrency.

Buying Cryptocurrency

Conceptually, to purchase and store cryptocurrency, you will:

1. Exchange fiat currency (such as USD) into a "primary" cryptocurrency such as Bitcoin (BTC), Bitcoin Cash (BCH), Ethereum (ETH), or Litecoin (LTC).

2. Transfer the previously purchased BTC, BCH, ETH, or LTC to a secondary exchange to trade for your choice of over 2,000 "smaller" cryptocurrencies.

3. Choose to either keep your newly purchased cryptocurrency on the exchange where it was purchased, or transfer it to one of the many cryptocurrency wallets available to investors for storing tokens.

The first step in purchasing cryptocurrency is setting up an account on a fiat cryptocurrency exchange, such as Coinbase.

Creating an account on Coinbase is quick and easy, and once you produce documents to verify your identity (so exchanges can comply with regulatory obligations) and are approved, you can purchase cryptocurrency via debit card or ACH transfer.

Coinbase is an Internet-based cryptocurrency exchange that allows investors to deposit fiat currency into an account to exchange for cryptocurrency. Additionally, Coinbase provides users with wallet addresses for all cryptocurrencies available on its platform.

Once signed up and inside your Coinbase account, navigate to the *Buy/Sell* tab.

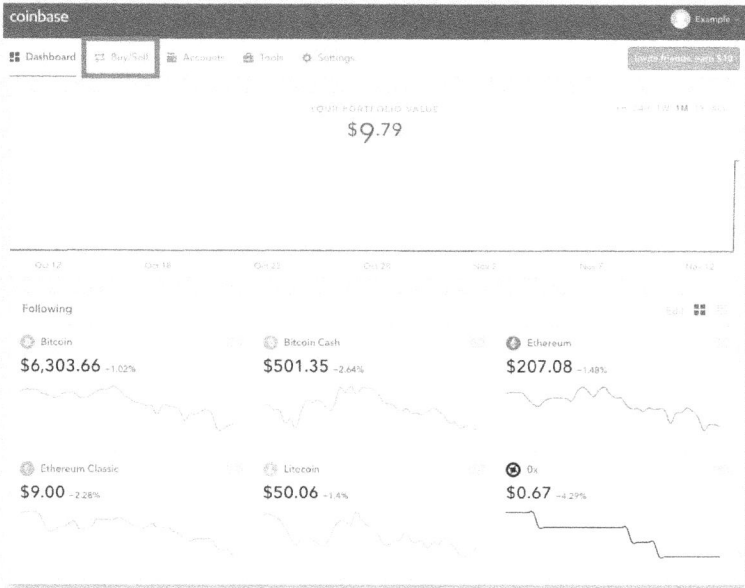

On the Buy/Sell page, select the coin you wish to purchase (from the drop down menu) and preferred payment method (debit card or linked checking account).

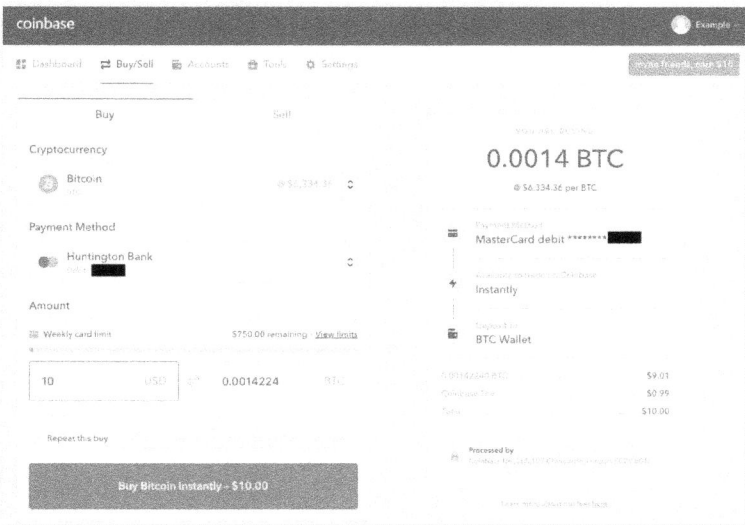

Confirm the purchase and navigate to the *Accounts* tab to view the coin deposited into your account.

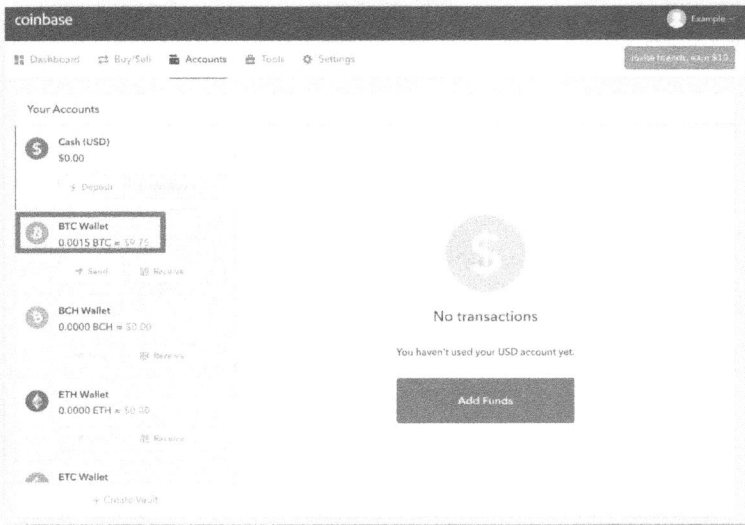

By clicking on the **Bitcoin** account, you can view your transaction history for the coin.

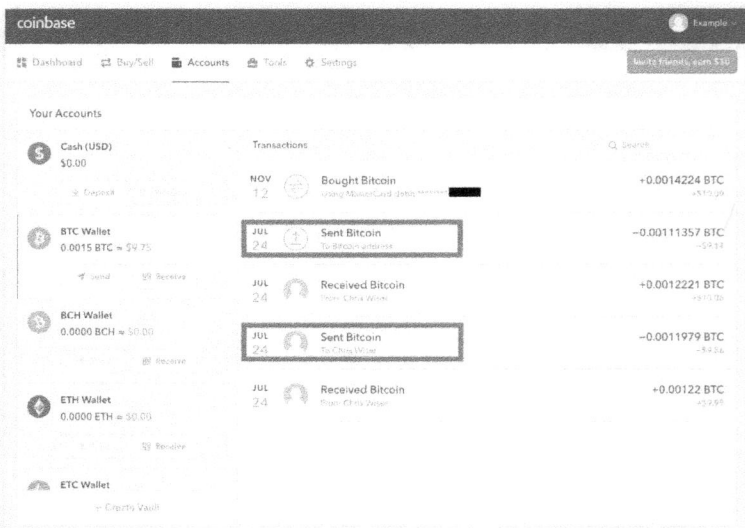

Note: Coinbase is unique in that it provides users the option to send and receive cryptocurrency via email address to other Coinbase users. However, if you elect to send cryptocurrency via email, your identity is no longer concealed under a public key, as shown above.

Coinbase—Sending and Receiving Cryptocurrency

The *Accounts* tab is also used to send and receive cryptocurrency.

Select **Send** to transfer cryptocurrency out of your account.

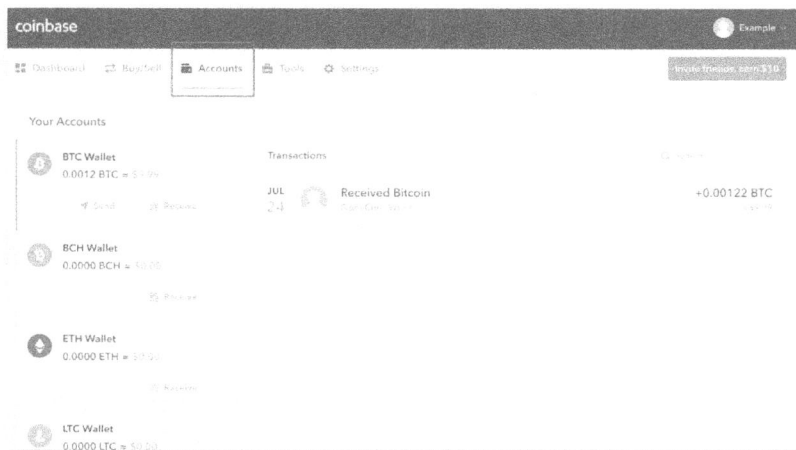

You will be prompted to enter the destination wallet address and amount.

Send BTC   ×

Wallet Address   Email Address

A miner fee will be added for sends to BTC addresses. Miner fees
do not go to Coinbase. To avoid miner fees, send to an email
address. Learn more.

Recipient

Enter a BTC address

Withdraw From

(B) BTC Wallet   0.00122000 BTC
                                             ≈ $9.97

Amount

0.00   USD   ⇌   0.00   BTC

Note

Write an optional message

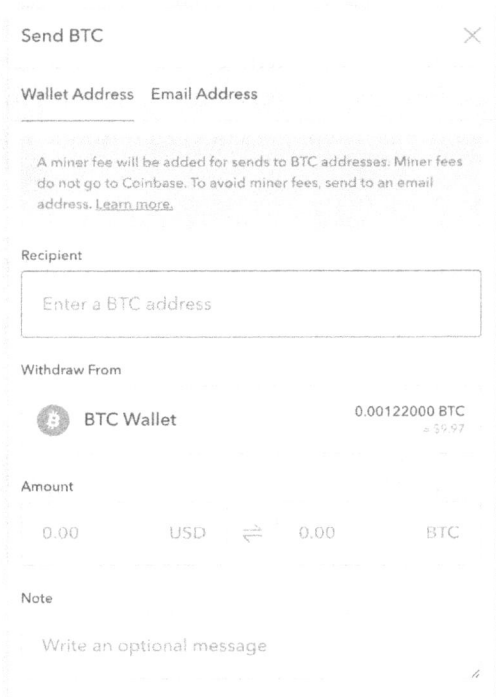

Click **Receive** to reveal your wallet address, which you will
provide to the individual sending you cryptocurrency. You will
also be shown a scannable QR version of your wallet address,
which can be used for error-proof and frictionless transfers via
some mobile applications.

BTC Wallet Address   ×

34edzJ2yqoDuF6nQbeVwuKAnZvFE83DPcM

Transfer from Coinbase to Smaller Exchange

To continue our example, we'll say you have purchased (or otherwise received) $10 worth of BTC into your Coinbase wallet.

If this is the cryptocurrency you ultimately want to keep, skip ahead to the next chapter on wallets to learn about the various options for storage and safekeeping.

If you wish to use this BTC to purchase another cryptocurrency, you must find an exchange where it is listed. This information is found on the company website, through a simple Google search, or with a resource such as CoinMarketCap.com.

We will use Binance, the popular cryptocurrency exchange, as an example interface, since it lists hundreds of "smaller" crypto-currencies and boasts features and navigation common to other secondary exchanges used to trade altcoins.

Upon navigating to Binance's home page, use the **Register** button to sign up. Similar to Coinbase, you will be required to verify your identity and setup two-factor authentication (2FA) to begin trading.

Once signed in to your account, you will be sent to a page with the header shown below:

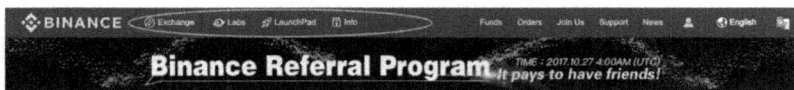

From this page, you are able to access all areas of your account by using the horizontal navigation bar.

Under the *Funds* tab, you are able to view balances by coin, as well as transfer money into and out of Binance.

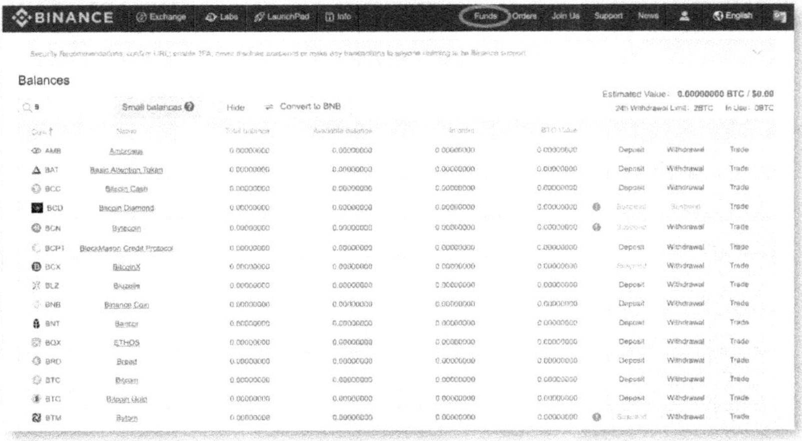

To transfer cryptocurrency into your account, find the coin of interest and select **Deposit**.

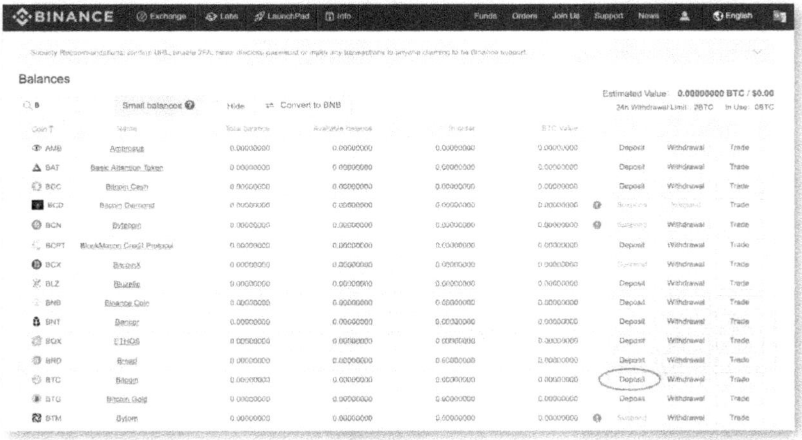

You will be taken to the following screen, which lists the BTC wallet address for your account.

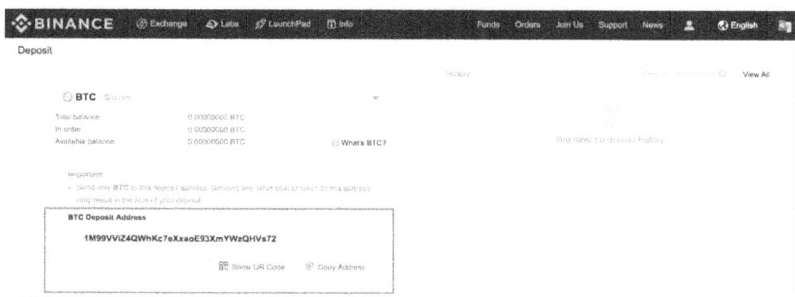

Note: If you have transaction history for this particular coin, transaction detail will appear in the right side of the screen, under *History*.

Select **Copy Address**.

Navigate to your Coinbase *Accounts* page, choose the cryptocurrency you would like to transfer to Binance, and select **Send**. When shown the following pop-up window, paste the BTC address from Binance.

You will know the address was pasted over correctly if you see the green check mark on the right side of the address box. However, it is always a good idea to double-check the address pasted into Coinbase to ensure it matches the wallet address you are looking to send coins.

Enter the amount you wish to send, in either BTC or USD, and select **Continue**.

Wallet Address   Email Address

A miner fee will be added for sends to BTC addresses. Miner fees do not go to Coinbase. To avoid miner fees, send to an email address. Learn more.

Recipient

1M99VViZ4QWhKc7eXxaoE93XmYWzQHVs⁊

Withdraw From

BTC Wallet   0.00124420 BTC
                                          ~$10.24

Amount

9.00   USD   ⇌   0.00109800   BTC

Note

Write an optional message

Continue

If you elect to set up the security setting, you must also enter a 2FA code before sending the transaction. Once completed, you will be given a confirmation screen with a summary of your transaction.

Sent Bitcoin ✕

↑

−0.00111357 BTC

≈ 59.14

To 1M99VViZ4QWhKc7eXxaoE93XmYWzQHVs72

Price per coin $8,207.84

Confirmations 0

Fee 0.00001557 BTC

Transaction View transaction

7/24/2018 3:49 PM PENDING

Note: By clicking **View Transaction**, you will be taken to a Block Explorer, which is a website that allows you to directly interact with the blockchain, check on the status of your transaction (as miners work to find the correct nonce), and research all past transactions. A primer for navigating public blockchain records via a Block Explorer is included near the end of this chapter.

Once your transaction is confirmed, the BTC will show up in your BTC wallet on Binance.

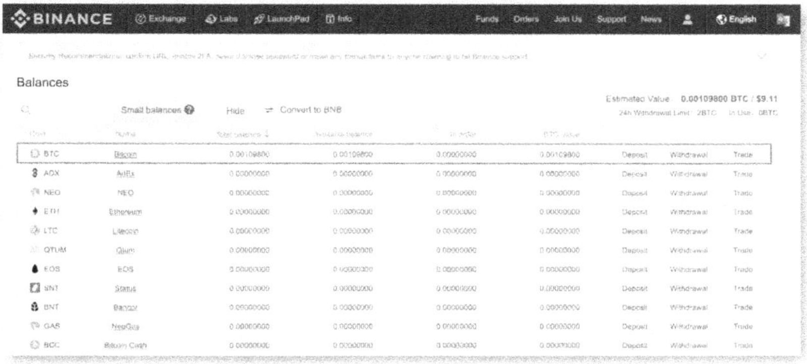

Next, we will look at how to trade the BTC deposited into Binance for your ultimate coin of interest.

A Primer and How-To Guide for Navigating Exchanges

To purchase cryptocurrency on Binance, navigate to the *Exchange* tab on the site's horizontal navigation bar.

On some exchanges, such as Binance, investors have the choice of basic or advanced trading screens. The basic option will likely suffice for most investors, but the advanced tab will give you the ability to analyze charts and use more in-depth trading tools, which are great features for active traders.

All the information seen on the exchange page is related to one specific trading pair.

To select the trading pair you are interested in, use the right side of the screen. We'll call this the "order desk," since it is the area of the site where you choose the cryptocurrency you would like to purchase.

Before choosing which cryptocurrency you would like to purchase on Coinbase, it is a good idea to identify one that can be traded into your ultimate token of interest. For example, if you ultimately want to buy XRP, you would want to make sure the token you purchase on Coinbase (BTC, in our example) can be traded to XRP.

By searching for "XRP" and toggling between **BTC** and **ETH** tabs at the top of the order desk, you can see that either coin can be traded into XRP.

★Favorites	BTC	**ETH**	**BNB**	**USDT**
XRP		Change		Volume
Pair ↑	Price			Change
★ **XRP/BTC**	0.00007153			0.15%

★Favorites	**BTC**	ETH	**BNB**	**USDT**
XRP		Change		Volume
Pair ↑	Price			Change
★ **XRP/ETH**	0.00119279			-1.18%

As you can see, the main cryptocurrencies used on Binance to trade into smaller currencies are Bitcoin (BTC), Ethereum (ETH), and Binance Coin (BNB).

Once you find the cryptocurrency pair you will trade, clicking on it will change all data on your screen to correspond with that pair. For our example, we will use the XRP/BTC currency pair.

Upon selecting this pair, several pieces of information become available to the user.

The most noticeable piece of information available is the price chart found in the middle of the screen.

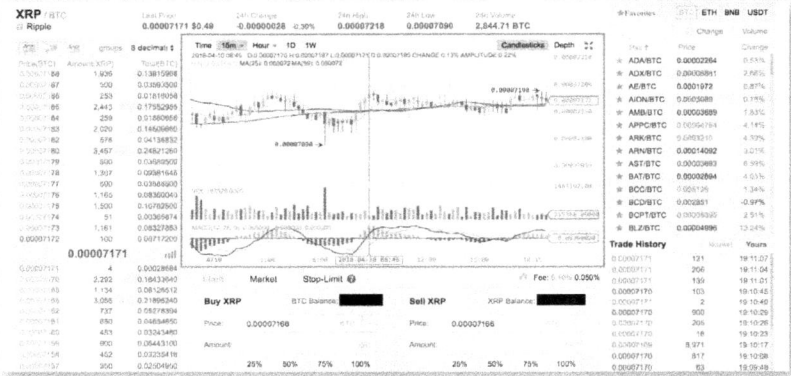

In this section, you can view the price history of a given token over various time horizons. In addition, you can overlay basic technical analysis indicators such as moving averages or volume traded in a given time period.

Just below the price chart section is the order ticket section, which is used to make trades.

Most exchanges provide a few trading options. Common options include:

- *Market Order:* An order placed to buy/sell an asset at the current market rate.

- *Limit Order:* A trade that will only execute when the price in the marketplace reaches a price that is better than or equal to a price you set. When buying, this can be used if you believe the current price of an asset is too high and you only want to purchase if the price comes down to a certain level; when selling, this is used to maximize profits by setting an exit point above the current market price.

- *Stop Loss Order:* A type of order used to curtail losses when selling an asset. A price lower than the current market price is set as your price floor, and if the asset drops to this price, your asset will be sold to prevent further losses. Stop loss orders are used to protect against large drops in the price of an asset you own. In this way, it *stops* your *losses*.

Let's look at an example:

In the screenshot below, you can see that the current market price for XRP/BTC is 0.00008137. That is, it takes 0.00008137 BTC to buy 1 XRP.

If you submit a *market order,* your trade will execute immediately at 0.00008137 BTC.

However, if you only wish to pay 0.00007 BTC for 1 XRP, you may place a *limit order* for this amount, and your order will not execute unless the market price drops to 0.00007 BTC/XRP — a better price than is currently offered in the marketplace.

Using a limit order to sell XRP is similar, and, in the same way you endeavored to pay the minimum when buying, you will look to receive the maximum when selling.

Therefore, you will place a *sell limit* order above the current market price, and your trade will only execute when the asset price appreciates to that level or higher.

However, there are risks in setting limit orders.

If you set a *buy limit* order (slightly below market price) and the market continues rising, you will miss out on that trade since the market price never dipped to fill your order. If you set a *sell limit* order (slightly above market price) too high, the market value may never reach that point before dropping and you could be stuck holding onto the asset as price continues to fall.

Even though you may be able to squeeze a few more dollars out of a trade when setting a limit order, it is slightly more risky because your target price may not be reached before the price of the security begins moving in the opposite direction.

Whereas limit orders are profit maximizing, stop loss orders boast a more protective function. When selling, you can set a stop loss to act as a "floor" price that, if the asset falls to, your trade will execute before the price slips lower. For example, if you purchase XRP/BTC at 0.00007, you may set a limit order at 0.00005 to limit your losses, should the trade turn against you.

Stop-limits allow you to feel comfortable investing in an asset as volatile as cryptocurrency without constantly checking the market.

In summary, a limit order is valuable because it allows you to define the price at which you are willing to buy or sell an asset to maximize profit, but gives you no protection from a price decreasing sharply in value. On the other hand, a stop-limit or stop loss order may be used to protect from the price slipping, but also allows you to enjoy the upside if the asset moves higher.

A market order is used when you aren't worried about that extra 0.0001137 BTC and just want to trade the darn thing!

As you can see, various order types may be appropriate depending on the purpose and strategy of the investor.

Due to the volatility of the market, learning to efficiently and properly use these various order types is crucial, especially if you plan on trading actively.

The next section, found on the left side of the screen, is known as the order book.

The order book shows a list of active orders in the market. This section is updated in real time as orders are created, canceled, and executed; therefore, this section may look entirely different every 5 seconds, especially if the cryptocurrency you're viewing has high trading volume.

The orders at the top of the section (in red) are sell orders. The orders at the bottom of the section (in green) are buy orders. This section has three columns:

1. *Price (BTC):* The price (in BTC) that the buyer/seller is willing to buy/sell XRP

2. *Amount (XRP):* The amount of XRP the buyer/seller is looking to buy/sell

3. *Total (BTC):* A multiplication of the previous two columns, which expresses how much the total trade is worth in BTC

You may have noticed that, in the **Price (BTC)** column, there is a range of prices that extend to each side of the current market price (circled above in blue). This is because, as referenced in the last section, traders may elect to only buy/sell at a specific price or better.

For example, at the top of the red section, we see one greedy seller wanting 0.00008180 BTC to part ways with their XRP, even though the market price is only 0.00008137 BTC!

There are ambitious buyers, too: the bottom of this section shows a buyer wanting to acquire XRP for a measly 0.00008129 BTC!

As demand for the XRP token fluctuates, the market will use this constant influx of buy and sell orders to negotiate a market price that strives to bring the demand to equilibrium.

In connection to the order book, trade history, which will be discussed in the next section, is found on the bottom right of the screen.

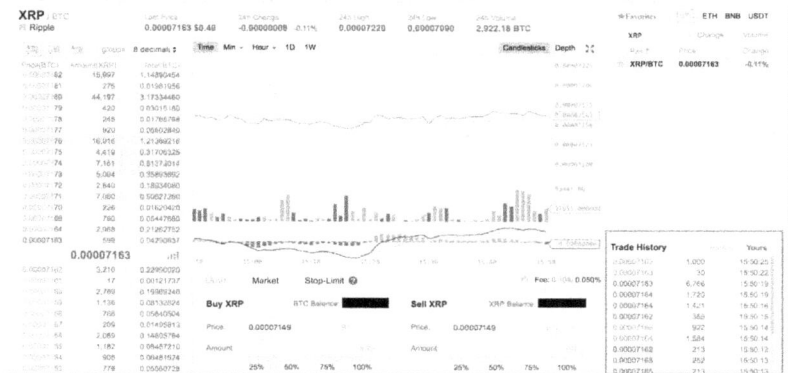

The trade history section simply displays the history of orders executed in the marketplace. The default option shows order history for all market participants; however, by toggling to the **Yours** tab, you can filter to only show orders placed by you.

Exchanges: Final Words

Now that you're oriented with the functionality of Binance, you can feel comfortable with other exchanges, too, since most navigation tools are similar across platforms.

Upon testing various exchanges, you will find that each platform has its own place in the cryptocurrency marketplace.

Some exchanges, such as Coinbase, only allow fiat-to-cryptocurrency trades and support a limited list of cryptocurrencies; whereas, exchanges such as Binance only permit crypto-to-crypto transactions but typically list dozens of cryptocurrencies. Therefore, you may need to utilize multiple exchanges to purchase your cryptocurrency of choice.

Coinbase has established itself as the favorite among retail investors and remains the preferred exchange for converting fiat currency to cryptocurrency, likely due to its user-friendliness. However, as the industry and marketplace progress, an increasing number of well known crypto-to-crypto exchanges are preparing to offer fiat-to-crypto trading, too. By consolidating these features under one platform and building out the user-friendliness needed to rival Coinbase, smaller exchanges are attempting to entice retail investors to make the switch from the long-time favorite.

ICOs

Initial coin offerings, or ICOs for short, are used by startups to issue shares of their business to interested investors through a token sale. Companies initiate ICOs to fund operating expenses while developing the application and getting off the ground. As we will explore in the next section, ICO campaigns are typically run during the earliest stage of a company's life, and since little information is available regarding past performance, ICOs are inherently high risk and unique due diligence is required.

If you are interested in ICO investing, you can use a service such as *icowatchlist.com* or *icoalert.com* to stay up-to-date on the newest projects.

The process for investing in an ICO is listed below:

1. Navigate to the company's website.

2. Peruse the site to learn about the team, application (through the technical white paper), roadmap for development, and token sale information. The site should also explain how the founding team would use the raised funds.

3. Sign up for an account and agree to terms and conditions, if necessary. If you are looking at an ICO that has not started, you have the option to get "whitelisted." Becoming whitelisted may permit earlier access to the sale or the ability to buy tokens at a discounted early bird price.

4. Once the ICO has begun, you are able to purchase the token. Many ICOs run on the ETH network. Therefore, you will

likely use ETH as the baseline token for exchange and an ETH wallet to receive and store your new ICO tokens.

a. Navigate to your wallet. Determine and enter the amount of ETH you wish to exchange and ensure you have at least that amount in your wallet.

b. The founding team will provide an ICO contract address, which is the wallet address where you will send ETH. Copy and paste this contract address into the *send* section of your ETH wallet.

c. Generate and confirm the transaction.

d. When the ICO sale ends, the smart contract governing the execution of the sale will automatically deposit the new tokens into your wallet at the predetermined exchange rate.

Note: It is very important that you DO NOT attempt to participate in an ICO with an exchange wallet address or other incompatible wallet, or you will be at risk of losing your funds altogether. Participating in an ICO requires you to send funds from a compatible, personal wallet. The ICO team will provide very clear instructions regarding the type of origination wallet required. If you have yet to create the required wallet type, you can find wallet selection and setup guidance in the next chapter.

The ICO company will likely also provide instructions for adding their token to your wallet for when it is deposited, but if not, refer to the section entitled "Setting up and Navigating a Wallet" in the next chapter.

Should I Invest in ICOs?

ICOs are the epitome of high risk, high reward.

Prior to 2018, there was virtually zero oversight or accountability for ICO campaigns. Anyone with an idea for a dApp could write a technical white paper outlining their project, build a website, and initiate a token sale to fund their endeavor.

Unfortunately, nefarious operators began capitalizing on the anonymous, remote, and underregulated nature of the marketplace. By creating a compelling website and white paper, these teams would mount aggressive marketing campaigns for a factitious product only to collect investors money, shut down the site, and disappear.

Unfortunately, recent research revealed that up to 70-80% of ICO campaigns run in 2017 were scams.

In response to these studies and the ICO mania of 2017, government agencies have initiated an onslaught of regulation in a strong response to this exploitation.

Still, ICO investing remains the most speculative option in cryptocurrency markets, since many ICOs represent companies that have yet to even begin building the product.

For these reasons, conducting due diligence is critical before investing in ICOs. In the next chapter, we will specifically focus on methods for evaluating ICOs that, if used properly, will help you spot the red flags and invest in projects with real potential.

Block Explorers: Behind the Scenes of the Blockchain

As stated earlier in the chapter, Block Explorers are used by participants in a network to directly interact with the blockchain and search the entire ledger of transactions, all the way back to the genesis block.

To use the Block Explorer select **View Transaction** on your transaction confirmation summary.

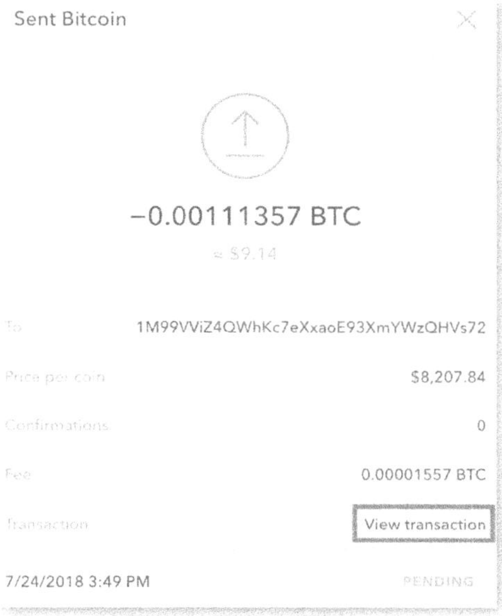

Sent Bitcoin ✕

↑

−0.00111357 BTC

≈ $9.14

To   1M99VViZ4QWhKc7eXxaoE93XmYWzQHVs72

Price per coin $8,207.84

Confirmations   0

Fee   0.00001557 BTC

Transaction   View transaction

7/24/2018 3:49 PM   PENDING

The Block Explorer you are linked to will provide information about the transaction, such as transaction number, amount sent, fees paid, block hash, and block number (shown below).

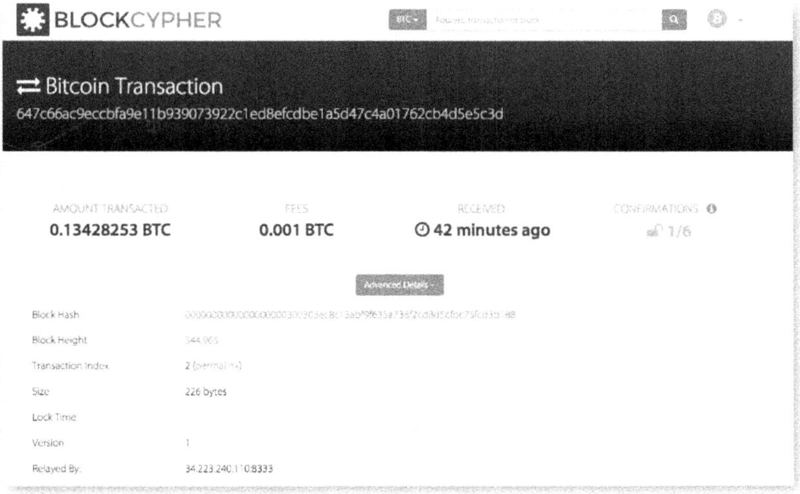

BLOCKCYPHER

⇄ Bitcoin Transaction
647c66ac9eccbfa9e11b939073922c1ed8efcdbe1a5d47c4a01762cb4d5e5c3d

AMOUNT TRANSACTED	FEES	RECEIVED	CONFIRMATIONS
0.13428253 BTC	0.001 BTC	⏱ 42 minutes ago	1/6

Advanced Details

Block Hash	
Block Height	
Transaction Index	2 (permalink)
Size	226 bytes
Lock Time	
Version	1
Relayed By	34.223.240.110.8333

Note: Coinbase currently uses BlockCypher (shown above), but the Block Explorer you are linked to will vary based on your exchange provider.

By refreshing the page, you are able to monitor the progress of your block being confirmed by miners and solidified onto the blockchain. In the example screenshot shown above, this transaction has achieved 1 of the 6 confirmations required to consider it complete and irreversible.

Once the Block Explorer indicates you have received sufficient block confirmations, the token will be deposited to your destination address.

By clicking any of the given hyperlinks on the page, you can learn more about this transaction.

For example, selecting **Block Hash** will reveal information about the block your transaction is located on, such as size of the block and details about other transactions it holds.

The alphanumeric string boxed in the following figure represents another transaction on your block.

By clicking on this transaction (which belongs to an unknown participant in the network), you become privy to all the details it contains, just as you would your own.

In this way, it is possible to sift through every transaction on the blockchain back to the genesis block.

Mining Pools

As discussed in Chapter 1, most cryptocurrency transactions require miners to use high-powered computers to solve computationally intense puzzles.

The amount of energy it takes to verify a transaction on any given network varies, but for a coin as energy dependent as Bitcoin, it is significant. For scope, the amount of power it takes to run the Bitcoin blockchain on any given day is equivalent to the consumption of some small countries.

Because the energy needed is so large, miners began combining their resources to solve blocks and split the block reward, leading to the creation of what we now know as mining pools. Many of the transactions sent today are on blocks solved by mining pools, since they have more horsepower than any single miner ever could. This is partially inherent to the natural maturation of the market, but many worry this could lead to the type of centralization of power that Bitcoin was created to oppose.

If a mining pool gained too much power (51%), it could theoretically work to overtake the network, begin rewriting history, and changing the data contained inside transactions. Since a key pillar of the first blockchain was decentralization, many critique blockchains with a high risk of this type of centralization.

Hard Forks

A fork occurs when an integral group of developers wishes to make a significant change to a blockchain's operating logic, but keep the core infrastructure. This change in operating logic forces a "fork" that leaves two resulting tokens: one that operates under the old set of rules, and an alternative, which operates under the new set of rules.

The resulting two coins represent two unique and independent blockchains, whose networks will cease interacting in any meaningful way. Nodes on the original blockchain no longer recognize transactions on the alternative blockchain, and vice versa.

Hard forks are often contentious by design, but remain a natural activity for driving iteration and development in the cryptocurrency ecosystem. Additionally, this open sourced, democratic governance ensures no group takes complete control over the direction of the project.

Notable examples of hard forks include:

- Bitcoin Cash, which changed the size of the block from 1MB to 8MB in an attempt to deal with scalability issues.
- Ethereum Classic, a fork of Ethereum, which was created following a hack on the Ethereum network.

Following any hard fork, both coins will press on to operate in accordance with their freshly contested values and operating system. Typically, if you own the coin being forked, you will be "airdropped" an equivalent amount of the new token into your wallet by the development team.

Transaction Times and Fees

Transaction times and fees will vary depending on the coin you are transferring and the wallets you are sending between. Some coins require the investor to pay a varying transaction fee to miners, and others don't require a fee at all. Similarly, each coin has an average transaction time depending on latency and inherent speed of the network. Websites such as *bitinfocharts.com* will show average transaction times, fees, and other transaction data by coin, so you can be prepared when sending between exchanges and wallets.

CHAPTER 5
Wallets: Storing and Sending Cryptocurrencies

Now, WITH CRYPTOCURRENCY successfully purchased and in hand, you have the option to either store it on the exchange or transfer it to a wallet that you — and you alone— own, control, and claim responsibility over.

A wallet, as mentioned previously, is a digital location where private and public key information is stored.

It is important to note that wallets do not store cryptocurrencies themselves, just the private and public key information. Wallets operate similar to accounts that allow the user to store, send, and transfer any other asset, such as a stock. However, while an account at your stockbroker is accessed through a central entity with a username and password, a cryptocurrency wallet allows the user to directly interface with the blockchain by "logging in" with a public and private key that are cryptographically linked.

Therefore, by using a cryptocurrency wallet to store and transfer cryptocurrency, you have more control over your assets, but also more responsibility. As outlined in Chapter 1, blockchain was created to enable the execution of transactions over an entirely decentralized, peer-to-peer network. While many tenants of an entirely decentralized, peer-to-peer system are exciting, this also places much more responsibility on the individual; there is no emailing customer support to reset your password if it is forgotten or misplaced. There are countless individuals who have lost significant sums of cryptocurrency because they misplaced their private key information and entirely lost access to their wallet.

With this shift of responsibility in mind, it is important to realize there are different types of wallets that all come with their own unique benefits and drawbacks. Choosing the wallet appropriate for you is a matter of understanding some of the trade-offs between different options.

Below is a guide to the various wallet types, their features, and relevant examples.

Wallet Options

All wallets can be grouped into two categories: *hot* and *cold.*

A *hot* wallet stores private keys in a location that is connected to the Internet, whereas a *cold* wallet stores private key information entirely offline.

Since hot wallets are connected to the Internet, they are generally more accessible, but come with the added risk of getting phished, hacked, or having your computer stolen and account compromised.

In the cryptocurrency ecosystem, many successful attacks have taken the form of nefarious websites phishing investors by duplicating the look and feel of well-known exchanges or hot wallets to steal login information.

Since hot wallets are connected to the Internet, they are subject to these phishing attacks, as well as more concentrated, targeted security breaches. For example, the infamous Mt. Gox security breach in 2014 was more of targeted cyber attack, versus a phishing scam.

Within hot and cold wallets there are more specific categories of wallets to choose from, each with their own benefits and drawbacks.

Desktop wallets are downloaded and installed on your computer from a software source whose code is publicly available and verified. These wallets store private key information on the desktop of your computer; therefore, they are very secure but could be compromised if your computer is hacked, stolen, or infected with a virus.

Examples: Bitcoin Core, Armory, and Exodus

Online wallets are accessed through a web browser, similar to an account at a bank. Upon visiting the site, a user logs in with a public and private key. Additionally, there are chrome extensions that act as online wallets and allow the user to interact directly with compatible sites (exchanges or Ecommerce sites), enabling truly frictionless payments. This storage method makes your assets very accessible but places them at a higher risk of phishing attempts and attacks since it is directly interfacing with the Internet.

Examples: MyEtherWallet, Blockchain.info, MetaMask (chrome extension)

Mobile wallets are accessed through a mobile application on your phone. The wallets are easy to use and work especially well for individuals that transfer cryptocurrency often, but can be compromised in similar ways as online wallets.

Examples: Coinomi and Jaxx

Hardware wallets store private key information on a small USB storage device. These devices store your private keys offline and can only be accessed from that device. Hardware wallets hold dozens of the most common cryptocurrencies, but not all of them, so it is a good idea to know the coin you plan to store before purchasing a hardware wallet. Hardware wallets are much safer than any hot wallet, but slightly less user-friendly and accessible; therefore, hardware wallets are often used to store large quantities of cryptocurrency that do not need to be accessed often.

Examples: Trezor and Nano Ledger

Paper wallets store your private key information entirely offline and on a physical piece of paper. Creating a paper wallet is the most extreme storage method and the most secure if appropriate precautions are taken. There are several services you can use to generate a paper wallet, but to achieve true confidence that keys are never transferred or stored on a server, you should use a generator that does not require a connection to the Internet and the code that governs the creation of the keys should be audited for integrity. The time and knowledge required to ensure the creation of a truly secure paper wallet is likely far too extreme for the average cryptocurrency investor; therefore, paper wallets are typically only used by investors storing a very large amount of assets.

Such extreme security measures may be justifiable when protecting significant wealth invested in cryptocurrency, but for 99% of cryptocurrency investors, paper wallets are likely not worth the hassle.

Examples: Bitcoinpaperwallet.org and Bitaddress.org

Part 1: Setting Up and Navigating a Wallet

Setting up a wallet is fairly straightforward; however, this section is dedicated to walking through wallet setup for the various types available to investors.

Online Wallets

We'll use MyEtherWallet (MEW), one of the most well known online wallets, as an example interface. This wallet is compatible with ETH, Ethereum Classic (ETC), and any coin built on the ERC-20 Protocol.

Before navigating inside the wallet, we will take note of a few key site features.

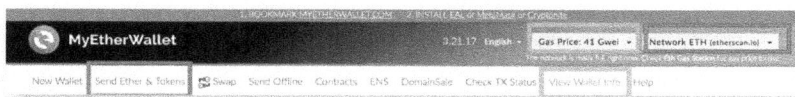

There are two drop-down menus on the top right side of the site. The box on the right indicates the blockchain **Network** you are interacting with — ETH or Ethererum Commonwealth (used for ETC). ETC is a fork of ETH and therefore shares some wallet compatibility. The functionality and navigation of both wallets is identical.

The box on the right shows the **Gas Price**, which is used to represent the current average transaction fee for Ethereum. In the Ethereum network, you have the freedom to set the maximum transaction fee (gas price) you are willing to pay, but beware: miners prioritize transactions they complete based on the highest reward (gas price) paid, so by offering a higher gas price, miners are incentivized to include your transaction in a block and solve it ASAP.

To send Ether and other ERC-20 tokens from this wallet, select **Send Ether & Tokens**. To sign into your wallet, select **View Wallet Info.**

However, before you can do either of these, you must follow the process below to create a new wallet.

Select **New Wallet** in the horizontal navigation bar. Choose a password, select **Create New Wallet,** and your private and public keys will be generated.

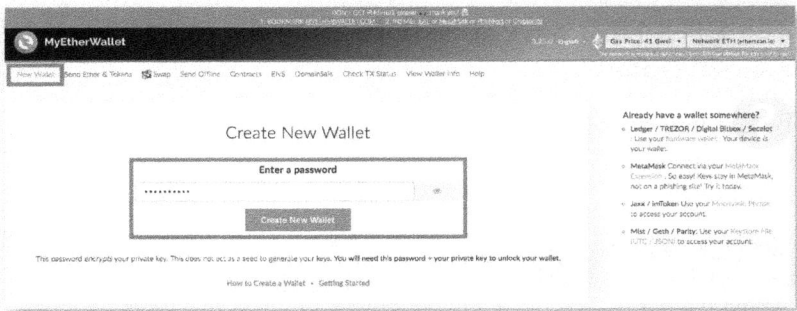

You will immediately be given the option to download a keystore file, which houses your newly generated keys and can be saved in any folder on your computer. This keystore file is one of a few ways to unlock your wallet, as we will see below.

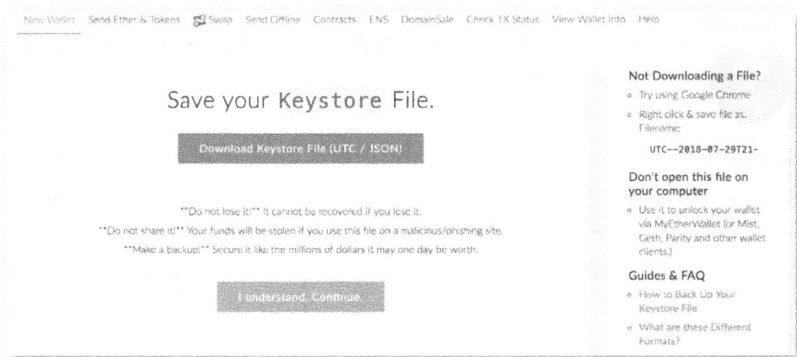

After selecting **Continue,** your private key will be revealed in text form, as seen below. Again, despite showing the private key for this example wallet, you want to keep this passcode secret and protected.

Save Your Private Key.

5e9a494b1a87f942e568abbbbd13a574547734b3bf90362bb39de2a9bb1f65c1

Print Paper Wallet

Do not lose it! It cannot be recovered if you lose it.
Do not share it! Your funds will be stolen if you use this file on a malicious/phishing site.
Make a backup! Secure it like the millions of dollars it may one day be worth.

Save Your Address. →

In some wallets, at this point, you are given a seed recovery phrase in case you misplace your private key. This recovery phrase consists of several random words and entering it is the only way to recover your keys. This recovery phrase should be stored in a safe location.

After saving your private key, you will be taken to the login page. Select one of the options listed to unlock your wallet.

New Wallet   Send Ether & Tokens   Swap   Send Offline   Contracts   ENS   DomainSale   Check TX Status   View Wallet Info   Help

Unlock your wallet to see your address

Your Address can also be known as you Account # or your Public Key. It is what you share with people so they can send you Ether or Tokens. Find the colorful address icon. Make sure it matches your paper wallet & whenever you enter your address somewhere.

How would you like to access your wallet?

MetaMask / Mist

Ledger Wallet

TREZOR

Digital Bitbox

Secalot

Keystore / JSON File

Mnemonic Phrase

Private Key

Parity Phrase

For example, if you choose to use a keystore file, you will be prompted to select a file from your desktop **and** enter your password to unlock the wallet.

Alternatively, you can simply elect to use the **Private Key** option.

If you own a hardware wallet, you may also access your assets through the MEW interface by selecting either **Ledger Wallet** or **Trezor**. After selecting this option and plugging in your USB device, you will be guided through unlocking your wallet on the screen.

During this login step, you may notice that MEW gives warnings about using a keystore file or private key to access your wallet. This is due to the security risks inherent to transmitting your private keys through a browser. For added security, store your assets in a Ledger or Trezor, in which your private keys never leave the USB device and MEW simply acts as an interface to the blockchain, allowing you to view and access the wallet contents.

Once your wallet is decrypted with the private key, you are able to see the inside, which looks similar to an account dashboard.

The two sections on the left side of your screen contain the private and public keys in different forms. The top box in the screenshot below shows your public and private keys in text form, whereas the fuzzy codes in the red box are scannable QR versions of these same keys.

The section to the right of these (at the top) shows wallet address, account balance, and a link to view the account transaction history and balance on Etherscan Block Explorer.

The section below account information (in the orange box) shows your non-Ethereum token balances, since most coins built on the ERC-20 protocol can be stored in this wallet, too.

By selecting **Show All Tokens**, you will see a list of all compatible ERC-20 tokens and their corresponding balances.

Token Balances

·̣̇· How to See Your Tokens

You can also view your Balances on
https://etherscan.io or ethplorer.io

Only Show Balances	Add Custom Token

0 $FFC
0 $FYX
0 $TEAK
0 0xBTC
0 1ST
0 1WO
0 2DC
0 300
0 A18
0 ABT

This section is especially helpful if you participate in an ICO and need to store an ERC-20 coin that may not have any other dedicated wallet option yet.

To find any new coin purchased through an ICO:

1. Select **Add Custom Token**.

2. Type in the contract address (the ICO address you sent funds to), the token symbol, and the decimals of precision (this will be given by the ICO founding team).

Token Balances

How to See Your Tokens

You can also view your Balances on
https://etherscan.io or ethplorer.io

Show All Tokens | **Add Custom Token**

Token Contract Address

Token Symbol

Decimals

Save

Equivalent Values

BTC: 0   REP: 0
USD: $0.00   EUR: €0.00
CHF: 0.00   GBP: £0.00

These are only the equivalent values for ETH, not
tokens. Sorry!

Once you select **Save**, the token balance will be added.

3. To find your token, select **Show All Tokens** again and scroll down to the symbol that represents the asset you purchased.

Note: If your balance remains 0, confirm the timeline for the company depositing funds after the ICO sale and ensure you are in the correct wallet. MyEtherWallet, for example, is an Ethereum and Ethereum Classic (ETC) wallet. If your wallet is toggled to **ETC**, your funds for an ICO supported by ETH wallets will not be visible. Switch to the **ETH** wallet to load your balances.

Remember, all transactions are visible to the public, so you can always easily search the transaction on a Block Explorer by using either the ICO contract address or your wallet address.

Part 2: Sending From a Wallet

To send from your wallet, select **Send Ether & Tokens** in the horizontal navigation bar of MEW.

On the next screen, simply enter the recipient address and amount to send. Press **Generate Transaction** and then **Send Transaction**. Once you confirm the transaction, use Etherscan to track it as it travels through cyberspace.

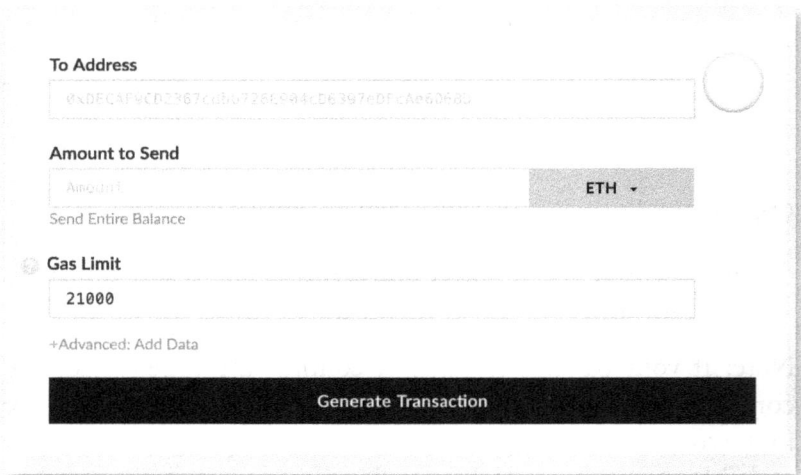

Mobile Wallets

Mobile wallets work similar to online wallets. Once the app is downloaded, log into your wallet to easily access and send cryptocurrency. By maintaining a mobile wallet of some kind, you are able to seamlessly send and receive cryptocurrency by scanning the QR code provided by other wallets.

Desktop Wallets

Desktop wallets are typically created by the developer of the token it stores, since lack of storage options could deter would-be investors from purchasing the coin.

However, multi-currency desktop wallets are oftentimes created by an independent organization that is unrelated to the coins it houses.

To create a desktop wallet:

1. Navigate to the company website to find instructions on downloading the wallet. The wallet download instructions will either direct you to a download page on the site or a public code repository, such as GitHub.

2. Download the file that contains the wallet to your desktop.

3. Follow the prompts to create a password and generate a seed.

4. Each wallet will look different, but all will have similar functionality to wallets explained earlier in the chapter. You will have a space to send, receive, and view the history of the wallet. There should also be a place to view the contents of your wallet and a scannable QR code.

Hardware Wallets

Hardware wallets can be ordered directly through a wallet manufacturer website or Ecommerce site. They will arrive as a USB device and contain very detailed instructions regarding set up. As mentioned above, hardware wallets hold your keys entirely offline but will use an online platform, such as MEW, to allow users to interface with the blockchain.

Paper Wallets

To **Deposit** funds into a paper wallet, simply provide the destination address listed on the paper to the sender. Additionally, most paper wallets have QR codes that can be scanned to easily receive cryptocurrency.

Withdrawing funds from a paper wallet should only be done once, since paper wallets are used for long-term storage and it is not considered safe to withdrawal partial balances.

To withdraw:

1. Choose a destination wallet that supports the ability to "sweep" paper wallets, such as Mycelium.

2. Depending on the service being used, you will select an option such as **Cold Storage** or **Import Paper Wallet.**

3. Scan the QR code, or enter the text form, of your paper wallet's *private* key.

4. Be sure to send the maximum amount from your paper wallet in order to entirely deplete its funds.

Again, paper wallets are meant for long-term storage, not daily transactions. While you can easily deposit funds multiple times, they should only be "swept" once, since the private keys are processed through a hot wallet during this "sweep."

Part 3: Choosing a Wallet

So which wallet option should you choose?

The answer to this will largely depend on your circumstance, but many investors maintain a variety of wallets. While an investor may store the bulk of assets to be held long term in a paper or hardware wallet, they will also likely store assets in an online or mobile wallet for easy access.

As discussed in the **Wallet Options** section, there are several trade-offs inherent to choosing a wallet type, and these trade-offs are best summarized by considering how each wallet ranks in the following categories:

- Responsibility
- Accessibility
- Convenience
- Control
- Security

The infographic below summarizes the pros and cons of each wallet type with the above criteria in mind.

WALLET TYPE	PROS	CONS
Dekstop Wallet	Readly accessible; medium security since private keys are stored on your computer, not online	High responsibility, hackable
Mobile Wallet	Frictionless access and transfer of coins	Security
Hardware Wallet	Secure and transportable; private keys stored offline	Expensive, low accessibility
Paper Wallet	Extremely Secure, total control over assets	Complex to set-up, can not be accessed quickly or frequently
Exchange Wallet	Seamlessy access and purchase new tokens	Security

With these tools and know-how as a foundation, we will now turn our attention towards how to evaluate cryptocurrencies for investment potential.

In the next chapter, we will walk through a step-by-step checklist and investing methodology that provides a framework for sizing up any given token.

CHAPTER 6
Cryptocurrency Investing Framework

IT's NO COINCIDENCE that *invest*ing and *invest*igate contain the same root word. You should never try to do the former without the latter.

In the cryptocurrency markets, this relationship is even more crucial.

Investing in cryptocurrency can be a bit less forgiving than other asset classes due to the volatility and uncertainty inherent to such an evolving market. In addition, cryptocurrency research looks much different than other investible asset classes. This is likely attributable to the lack of regulation, risk of fraudulent activity, and fast pace of the market.

In this chapter, we will explore exactly how to conduct the critical due diligence required when investing in cryptocurrency markets.

Researching Cryptocurrency: Your Checklist

What makes a coin go up or down? What makes it succeed in the long run?

There are general rules to help answer these questions in other asset classes such as stocks and bonds, but cryptocurrency is an entirely new class with entirely different rules. Below is a checklist for topics you should research when considering whether or not to make a specific cryptocurrency purchase.

Use Case(s)

What is the problem the project is looking to solve? What is the case, or scenario, in which this proves useful? What *exactly* is the company creating and why? Does the application sound far-fetched or overly niche? To gain the best understanding of the use case and its relevance, read the company's white paper, as it remains one of the best ways to become familiar with its application, vision, and roadmap. However, it is important to understand that the company's white paper is a forward-looking statement. For this reason, you should consider development of the application over time, if possible. Since the project's inception, has the founding team stayed on track with development and made significant progress, or gone dark with minimal communication or development updates?

Also, make sure you understand (and believe in) the value proposition for adoption by an individual or enterprise. Heed the advice of Warren Buffett on this and "never invest in a business you cannot understand."

Performance of Blockchain

Ability of the blockchain software to effectively perform how the team proposes is one of the most important areas for research. If the purpose of the project is to offer a universal payments system, it will not prove effective if it has a maximum capacity of 1 transaction per second (Mastercard can process around 2,000).

The same is true for all blockchain applications: you must make sure the software infrastructure is robust and performs well enough to justify a use case to the would-be end user.

Prevailing Interest in Application

What is the current interest in the company? Do they solve a substantial problem for enterprise users, governments, or end users? If not, their business may not be likely to succeed since their product won't be widely adopted.

By researching the target market of a blockchain application, you gain insight into its growth potential.

For example, asset digitization is a blockchain application likely to be adopted in the coming years due to the ever-expanding need for shipping and tracking goods. However, this does not mean all companies building asset digitization blockchains will be successful, just that this category could have a great deal of potential due to prevailing societal needs.

Partnerships

This is similar to understanding the end user interest in the application. One indication of a potentially promising blockchain project is strong partnerships with existing organizations that operate adjacent to their application. Strong partnerships indicate a proven entity is supportive of a project and the team behind it; it may also indicate that entity as a future investor or end user of the technology.

Partnerships have proven to be a big deal for organizations creating dApps. For example, in 2017, the IOTA Foundation announced a partnership with a handful of household name enterprise users, such as Microsoft and Accenture, and saw a sharp increase in their token price as a result. Shortly after, they clarified that no formal partnership was initiated but, rather, the aforementioned companies agreed to be participants in the IOTA Foundation's data marketplace pilot. Even though this still proved a significant breakthrough, their founding team faced backlash and the token saw a noticeable deprecation in value in light of the previous announcement.

For these reasons, investors should keep an eye on blockchain applications that larger corporations are investing in, piloting, or integrating. This attention may influence that project (and the attached cryptocurrency) being adopted and successful in the long term.

Team

What is the track record of the team building the dApp? The credibility and horsepower of the development team is crucial, since they are building cutting-edge software; however, good leadership from CEO's, advisors, and other strategic roles is crucial to developing partnerships, ensuring adequate funding, and managing the project timeline.

Often times, integral members of a successful and established token become members of the Executive Board or founding team of lesser-known projects. Having someone with a proven track record in the space can be an asset for a new project, both strategically and operationally. Look into the team of the token you are interested in; you can fairly quickly find either red flags or the validation needed to consider the project further.

Community and Communication

The overall community interaction surrounding a given token is important to take into consideration. Communication mediums such as Slack, Telegram, Discord, and Twitter sometimes enable direct interaction with the team and allow you to sense the general posture of the conversation surrounding the project. A good team may provide frequent updates on progress, announce new partnerships, showcase pilot test results, or comment on the landscape of their application. In 2017, ICO scams would have been much less fruitful if investors would have simply recognized red flags found in communication channels, which revealed the absence of a legitimate and responsive team.

Manipulation

If the founding team keeps a large percentage of tokens issued, either as an investment or stored in a reserve, they could retain the ability to manipulate the price of the coin in the marketplace by buying or selling large quantities.

In the stock market, there are measures in place to combat selling by founders (holding periods, insider trading announcements, etc.), but in the underregulated world of cryptocurrency, there is no limit to the amount of tokens that can be sold at any time by anyone.

This type of underregulation also leads to security fraud schemes such as the "pump and dump." In these schemes, unethical founding teams and/or other investors have been caught paying influential figures to promote their coin to *pump* the price up. Once the price is inflated, actors involved in the pump begin selling their coins for profit. This *dump* of sell orders then drives the price down and other investors are stuck holding a token that is worth much less than when they bought it. Following 2017, several public figures came under scrutiny or investigation for participating in these pump and dump promotion schemes.

Someone investing for the long(er) term need not learn to analyze charts to the same degree as more active traders, but learning to recognize a pump and dump may help ensure that a coin you are looking to invest in is built on good foundation, not hype.

These various forms of market manipulation are the subject of more interest and investigation by regulatory agencies such as the SEC, FTC, and CFTC. This governance, as well as increased regulation of ICOs, will continue to weed out the illegitimate actors that initiate these activities.

Still, while this regulation is in its infancy, and even after it is fully developed, it is important to remain vigilant enough to spot and protect yourself from these scams.

Economics and ICO Practices

Even if the project you are interested in had their ICO several months or years ago, there are a few key decisions made during an ICO sale that investors should note.

Supply: How many tokens were issued in the token sale? Some teams attempt to artificially bolster their market capitalization (# of coins released x price per coins) by releasing an excessive supply.

If a company issues 1 trillion coins in their token sale (yes, it has happened), each token is likely to become diluted in value, in order for the market capitalization to reflect true market value.

These coins will likely be traded at a fraction of a penny, which is unattractive to investors due to poor liquidity and increased volatility, making the token unlikely to appreciate in value.

Distribution: Once you know how many total tokens are being created, you need to know *how* the tokens are being distributed. The founding team should outline both how the tokens will be divvied between investors, founding team, and partners as well as how the raised money will be spent (development, advertising, legal expenses, etc.). This ICO information should be clearly listed on the company website.

As an investor, it is important to agree with how raised funds will be spent. Consider that allocation for advertising and sales will enable the company to drive adoption, but that is no substitute for development and testing of the actual application.

Valuation and Issuing Price: By calculating how many tokens you are receiving for a given amount of ETH (or whichever token you are exchanging), you can determine what the team sees as the value of each token. By multiplying this by the amount of tokens issued, you can determine the price valuation given to the company by the founding team. Do you agree with this valuation given your research and in benchmarking it against similar projects?

Team Members: For the reasons listed above, it is a good idea to see that members of the founding team have experience in finance, economics, ICO funding and strategy. Additionally, the team must be built on a strong technical foundation to support the protocol or application being built.

Previous Funding

For an ICO, this isn't crucial, since the team is using the ICO to crowd source funding. However, seeing previous investment from a venture capital fund or well-known angel investor could be a positive sign for an early stage ICO. For tokens past the ICO stage (currently trading on the exchanges), investment from enterprise partners or institutional investors can denote heightened interest.

Accessibility and Liquidity

Which exchange(s) is the token listed on, if any? This is especially important when investing in ICOs. It is always a good sign to see new projects displaying a partnership to be listed on an exchange directly following their ICO, as this speaks to interest in, and legitimacy of, an application. Also, if investors have limited access to trading the coin, the token is not likely to appreciate in value, sometimes regardless of the underlying blockchain's performance.

Research Tools

With these research topics in mind, every investor must also become comfortable with navigating and properly utilizing the ever-increasing number of cryptocurrency research repositories available.

Among the most foundational and long-standing sites for crypto-currency updates and research is *Coinmarketcap.com*, which provides a wealth of information on tokens, exchanges, and the overall market.

Initial Research—CoinMarketCap.com

Navigate to Coinmarketcap.com. On the main page (shown below), you will see a snapshot of the market. At the top of the page, you will see a thin bar (in the pink circle) that tracks the total number of cryptocurrencies available for purchase, total volume, and total market capitalization. By using the navigation bar found in the green circle, you are able to search currencies and navigate to pages that provide high-level market data.

Top 100 Cryptocurrencies by Market Capitalization

Select the cryptocurrency of interest. We will use **Litecoin** as an example.

Top 100 Cryptocurrencies by Market Capitalization

This will take you to a page that displays the price history of Litecoin in chart form and other market metrics related to the token.

Along the top, you will find information such as market capitalization, trading volume, and circulating supply. You can also see the price expressed in relationship to Bitcoin. Bitcoin is commonly used as the crypto-to-crypto measuring stick for valuing currencies; however, most cryptocurrencies also express their price in relation to the USD.

On the left side of the screen, there are links to the website, recent announcements, and even the source code that enables its blockchain to operate.

In the middle of the screen, you will find a few tabs to toggle between, such as markets, social (media), tools, and historical data.

The *Markets* tab is most commonly used and is explored below.

Selecting *Markets* will take you to a page that provides exchange-focused information, such as the leading exchange in Litecoin trading volume and which exchanges can be used to purchase this cryptocurrency. For example, the screenshot below shows that the LTC/BTC pair on Binance is the 5th most traded Litecoin pair across the entire marketplace.

On this page you, will notice slight price discrepancies between exchanges. However, it is important to note that these price discrepancies are proportional. For example, if Litecoin is more expensive on exchange A, all other cryptocurrencies on exchange A will usually be proportionally more expensive. Therefore, the exchange rate for Litecoin to any other cryptocurrency between exchanges will be nearly identical.

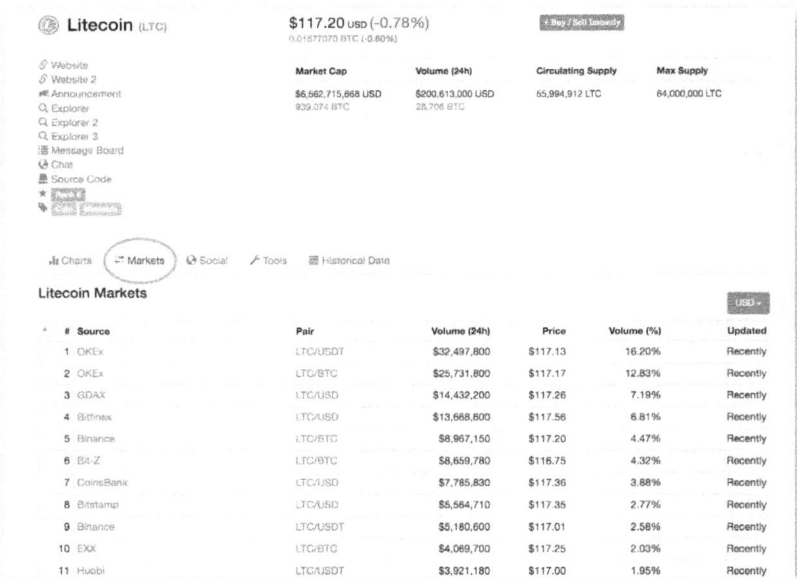

Litecoin (LTC) $117.20 USD (-0.78%)

	Market Cap	Volume (24h)	Circulating Supply	Max Supply
	$6,562,715,868 USD	$200,613,000 USD	55,994,912 LTC	84,000,000 LTC

Litecoin Markets

#	Source	Pair	Volume (24h)	Price	Volume (%)	Updated
1	OKEx	LTC/USDT	$32,497,800	$117.13	16.20%	Recently
2	OKEx	LTC/BTC	$25,731,800	$117.17	12.83%	Recently
3	GDAX	LTC/USD	$14,432,200	$117.26	7.19%	Recently
4	Bitfinex	LTC/USD	$13,668,800	$117.56	6.81%	Recently
5	Binance	LTC/BTC	$8,967,150	$117.20	4.47%	Recently
6	Bit-Z	LTC/BTC	$8,659,780	$116.75	4.32%	Recently
7	CoinsBank	LTC/USD	$7,785,830	$117.36	3.88%	Recently
8	Bitstamp	LTC/USD	$5,564,710	$117.35	2.77%	Recently
9	Binance	LTC/USDT	$5,180,600	$117.01	2.58%	Recently
10	EXX	LTC/BTC	$4,069,700	$117.25	2.03%	Recently
11	Huobi	LTC/USDT	$3,921,180	$117.00	1.95%	Recently

To view all relevant information about the platform, click on the name of any exchange. We will use Bitfinex as an example.

🍥 **Litecoin** (LTC) $117.20 USD (-0.78%)
0.0167707 BTC (-0.80%)

[+ Buy / Sell Instantly]

	Market Cap	Volume (24h)	Circulating Supply	Max Supply
🔗 Website	$6,562,715,668 USD	$200,613,000 USD	55,994,912 LTC	84,000,000 LTC
🔗 Website 2	939,074 BTC	26,706,91C		
📢 Announcement				
🔍 Explorer				
🔍 Explorer 2				
🔍 Explorer 3				
📋 Message Board				
💬 Chat				
🗂 Source Code				
★ [Rank 6]				
🔖				

📊 Charts   ↗ Markets   🌐 Social   🛠 Tools   📅 Historical Data

Litecoin Markets

[USD ▾]

#	Source	Pair	Volume (24h)	Price	Volume (%)	Updated
1	OKEx	LTC/USDT	$32,497,800	$117.13	16.20%	Recently
2	OKEx	LTC/BTC	$25,731,800	$117.17	12.83%	Recently
3	GDAX	LTC/USD	$14,432,200	$117.26	7.19%	Recently
4	Bitfinex	LTC/USD	$13,668,600	$117.56	6.81%	Recently
5	Binance	LTC/BTC	$8,967,150	$117.20	4.47%	Recently
6	Bit-Z	LTC/BTC	$8,659,780	$116.75	4.32%	Recently
7	CoinsBank	LTC/USD	$7,785,830	$117.36	3.88%	Recently
8	Bitstamp	LTC/USD	$5,564,710	$117.35	2.77%	Recently
9	Binance	LTC/USDT	$5,180,600	$117.01	2.58%	Recently
10	EXX	LTC/BTC	$4,069,700	$117.25	2.03%	Recently
11	Huobi	LTC/USDT	$3,921,180	$117.00	1.95%	Recently

By clicking on **Bitfinex**, you can see a list of all cryptocurrencies being traded on the exchange, volumes, prices, and currency pair options for trading.

Bitfinex $296,494,736 USD
42,218 BTC

🔗 https://www.bitfinex.com
🐦 @bitfinex
🔖 [merchant]

Active Markets

[USD ▾]

#	Currency	Pair	Volume (24h)	Price	Volume (%)	Updated
1	Bitcoin	BTC/USD	$177,524,000	$7,017.80	59.87%	Recently
2	Ethereum	ETH/USD	$39,809,200	$398.59	13.43%	Recently
3	Litecoin	LTC/USD	$13,596,700	$117.61	4.59%	Recently
4	EOS	EOS/USD	$10,562,400	$5.95	3.56%	Recently
5	Ripple	XRP/USD	$8,724,840	$0.495040	2.94%	Recently
6	Bitcoin	BTC/EUR	$5,717,370	$7,024.34	1.93%	Recently
7	Ethereum	ETH/BTC	$4,684,210	$398.44	1.58%	Recently
8	IOTA	MIOTA/USD	$4,560,100	$1.02	1.54%	Recently
9	Bitcoin Cash	BCH/USD	$4,338,690	$654.09	1.46%	Recently
10	Ethereum Classic	ETC/USD	$4,045,380	$14.15	1.36%	Recently
11	NEO	NEO/USD	$2,873,100	$48.08	0.97%	Recently
12	EOS	EOS/ETH	$2,257,210	$5.96	0.76%	Recently
13	EOS	EOS/BTC	$1,840,440	$5.96	0.62%	Recently
14	Dash	DASH/USD	$1,787,580	$314.44	0.60%	Recently
15	Ripple	XRP/BTC	$1,591,580	$0.494788	0.54%	Recently
16	Monero	XMR/USD	$1,334,820	$172.20	0.45%	Recently
17	Bitcoin Cash	BCH/BTC	$1,248,220	$652.36	0.42%	Recently

By using the drop down menu on the right side of the page, you can toggle between prices reflected in USD, BTC, ETH and many other currency denominations.

Supplemental Resources

To accomplish the aforementioned analysis, there are several tools at your disposal to aid you in your efforts. CoinMarketCap is a worthwhile starting point and quick dashboard, but only the tip of the iceberg.

Today, the resource list for sites that provide cryptocurrency-related information can seem as ever changing as the market itself. However, the list below provides examples of a few integral news outlets and other research sites that may come in handy.

- *Coincheckup*: A true one-stop-shop for cryptocurrency research

- *Cointelegraph*: News, price analysis, market tools, and even blockchain job openings

- *Cryptocompare*: A nexus for wallet and exchange reviews, price updates, and news

- *Telegram:* Channel-based messaging platform; a staple in the cryptocurrency world for following the inner circle of a project, especially during grassroots stages

- *Medium:* News site with active cryptocurrency and blockchain channels, featuring quality publications from a multitude of both individual contributors and organizations

- *Investopedia:* A vault for all investing and finance education

Handling Risk When Investing

As you likely already know—and has been frequently mentioned—investing in cryptocurrency is inherently risky, due to the immaturity of the market, question marks in its regulatory landscape, volatility, and the overall esoteric nature of the security.

Remember that all cryptocurrencies, especially those in the ICO stage, represent very early stage companies. Even some of the most well-known and established tokens would still be most appropriately classified as startups in the larger business community. This is partially what makes the space so exciting, but also what makes it inherently risky.

Risk management is especially important when investing in cryptocurrency. All of the normal rules you would apply to investing still apply here:

- *Risk Threshold:* By nature, cryptocurrency investing is high-risk, high-reward and very volatile. If you are going to invest, you must understand that it has proven to be a very fast moving market, so be sure you can stomach the ups and downs.

- *Diversify:* Diversification will allow you to capture the overall gains of the market even when a few particular projects are not performing well.

- *Portfolio Allocation:* Because of the aforementioned risk, it should be obvious not to take all your money out of stocks, bonds and more stable investments to put into cryptocurrency. For reference, commonly formed opinion for asset allocation into cryptocurrency is shown below:

Crypto Asset: Portfolio Allocation	
MAX % of portfolio in cryptos*	Experience Level
1.00%	Low
2.00%	Low to Moderate
3.00%	Moderate
4.00%	Moderate to Advanced
5.00%	Advanced
*Subject to market view	

Final Words

As the industry has developed, we've seen these project groups transform from ragtag groups of programmers to legitimate and investible startups. This has generally produced more data, which is now available for investors to use in researching tokens.

Remember, investing in a cryptocurrency is investing in a company—and all the ideas, strategy, and operations that will help it either fail or succeed.

The checklist found in this chapter contains information to consider—and resources to utilize—when researching cryptocurrencies.

With this checklist in mind, you should feel more comfortable evaluating the potential of any given cryptocurrency in the marketplace. By utilizing this research style and tools provided in previous chapters, you now have the resources necessary to safely navigate the cryptocurrency ecosystem. In the final chapter, we will tie together everything learned up to this point by both examining the current state of the industry and discussing its possible future direction.

CHAPTER 7

The Future of Blockchain and Cryptocurrency

HOPEFULLY, THIS BOOK has proven effective as an exhaustive primer and resource guide, and helped answer important questions about cryptocurrency and blockchain, such as:

- What are they?
- What is their relevance?
- How could they affect me?
- How should I interact with them?
- What should I look for when considering investing?

In this chapter, we will evaluate the current state of blockchain and cryptocurrency and discuss major developments expected as adoption continues.

Blockchain: A State of the Union

2017 and 2018 were exciting years for blockchain, especially.

In a relatively short period of time, the organizations building blockchain applications have made impressive progress and the technology has reached ubiquity across many industries and sectors. Today, there is undeniable interest, money, and support from investors, multinational corporations, and governments across the globe to further the development of this new technology.

This support has facilitated the creation of powerful partnerships and research labs that are focused on pilot testing various blockchain applications.

Through these pilot tests, startups have continued to refine their offerings while allowing their corporate partners to see how blockchain may fit into, and improve, their operations.

Below are a few examples of current and ongoing blockchain pilots:

- Fintech startups, such as Ripple, are piloting applications to streamline cross-border payments for a select group of industry partners. The results of these pilots are promising, as Ripple proved effective in facilitating international payments at a fraction of the time and cost of other services. In an increasingly global world, it is easy to imagine how such a use case could revolutionize the cost and latency associated with international commerce.

- Wal-Mart, along with a consortium of other retailers, has been collaborating with IBM since 2016 to create a blockchain that increases visibility to food as it travels through the supply chain. The testing began in 2017 and, since then, Wal-Mart has shared significant progress, including shortening the time required to track the lifecycle of produce from 7 days to 3 seconds. These types of improvements in data visibility could have a marked impact on recall management, food safety, and waste reduction.

- The immutability feature of blockchain technology has been popular in building blockchain-based voting platforms. Companies using proxy voting and municipalities that poll their constituents are interested in how blockchain-based voting systems can improve absentee ballot security and effectiveness, remove election fraud, and provide immutable voting records. Recently, entities ranging from multinational corporations to city governments have announced pilots for this purpose in blockchain hubs.

To support this increase in demand, the job market for blockchain developers and integrators has exploded, with many banks and businesses creating entire departments or teams dedicated to blockchain research and integration. To educate this new workforce, universities and online course libraries are racing to add classes, certifications, and degrees in blockchain. Blockchain events and conferences are on the rise and continue to pull experts from around the globe and boast household name sponsors, with some of the world's largest and most technologically advanced companies attending in an attempt to understand and leverage the quickly developing technology.

By many indications, blockchain is on a healthy path of development and maturation.

With that said, the technology still has plenty of inherent risks.

At its core, blockchain is proposing an entirely new system for executing transactions and storing data.

Some of the applications being built (banking, privacy, government record keeping, etc.) propose an overhaul to systems that protect and govern very sensitive information. While the security protocol improvements proposed by blockchain are impressive, and many technologists are supporting it's possible uses, nothing is unhackable. As an immature technology, there is plenty that still needs testing.

Fortunately, there is ample effort by developers, integrators, and practitioners dedicated to stress testing these applications.

Why Blockchain? Why Now?

Throughout history, each new technological innovation ultimately paved the way for several more to follow. This new invention may fossilize its predecessor, or it may simply parallel, support, accessorize, or augment the original.

Let's consider computers as an example.

The first computers provided a noticeable improvement to the paper-and-pen systems previously used for performing computations and storing data and files. Eventually, we were able to make these computers communicate with each other, thus enabling the transmission of computations and records between nodes.

Enter the World Wide Web (Internet) and email.

These two inventions entirely revolutionized our ability to connect, access, and share information.

Shortly following, we experienced a very wide tree of inventions aimed at accessorizing and supporting the Internet. These inventions include increasingly portable and powerful digital computers, Ecommerce, cloud storage, cybersecurity, and many more. Each of these innovations solved a new problem as computers developed and our needs changed.

As civilization progresses, opportunities are revealed and the next logical step for an innovation becomes clear.

When considering an innovation such as blockchain, it is helpful to ask, "What developing need does this satisfy?"

Recently, a large societal need has been found in the exponential growth of data generation and use. In an increasingly connected and digital world, we have an ever-increasing amount of devices and sensitive information that require safekeeping and management. Today, so much data is generated daily that organizations and individuals pay just to store it, let alone process it to unearth insights!

In short, the growth of data generation has far outpaced our ability to store and process it.

Enter blockchain.

Many regard blockchain as a contender for the next step in database management and transaction processing, due to its potentially improved storage and autonomous transaction mechanisms. In this way, blockchain could be viewed as a much needed upgrade to these areas.

Of course, the promise of blockchain is continually dependent on its development, testing, and adoption by end users.

At this point in blockchain's development, anything is possible— from discovering it is the new standard for data processing, to uncovering a fatal flaw in the technology rendering it unusable on a large scale.

If blockchain continues to perform well and proves a feasible update to current systems, it would be of interest to any organization wishing to secure, transfer, or store data. In an increasingly data-driven world, this could mean a slew of potential end users for the technology.

Since this book can only act as a static foundation and introduction to this new technology, and due to the incredibly fast pace of the industry, it's now up to you to stay on top of the ongoing progress in the space.

By using the applications discussed in Chapter 2 as a starting point, and some of the tools mentioned in previous chapters, take time to research and make a judgment on which categories of applications—or specific businesses—you believe will flourish and become pioneers in the blockchain industry.

Blockchain: Final Words

Blockchain has quickly transformed from an unknown or questionable concept to a disruptive emerging technology that boasts the potential to revolutionize data storage and processing. By allowing the world to digitize and streamline the execution of any transaction, blockchain applications have potential in sectors such as finance, banking, healthcare, logistics, government, retail, law, farming, and many others.

At this point in time, the potential utility of blockchain is embraced, or at least respected, by most who understand it.

Cryptocurrency: A State of the Union

Theoretically, the development of any given cryptocurrency should mirror the development of its associated blockchain application.

Unfortunately, this hasn't been the case: while the conversation surrounding blockchain seems to stabilize daily, the prevailing opinion of cryptocurrency remains largely polarized.

And why wouldn't it be?

The anonymous and mysterious nature of cryptocurrency encouraged its initial adoption in black market transactions and money laundering activities. In addition, we've discovered notable market manipulation, tax evasion, and general fraud connected to trading and ICO campaigns.

In hindsight, these outcomes shouldn't be surprising.

Cryptocurrency—as a financial instrument—and the market-places that support it, were created to be free of all the regulation enacted in direct response to these known illicit activities in capital markets. While the hallmark purpose of cryptocurrency was to anonymize and decentralize transactions, an unintentional byproduct was removing the governance historically relied upon to combat these unsavory activities.

Many of these issues are encouraged or amplified by how fast cryptocurrency grew in 2017. As new entrants poured in, the markets grew at blazing speed, and everyone scrambled to keep up. Exchanges, wallet providers, and other firms simply did not have the infrastructure to support the exploding demand, and customer service suffered. In addition, unbearable network latency and unstable platforms led to constant complaints of money being difficult to access or lost altogether.

To compound this problem, uneducated investors firing from the hip without doing adequate research made the cryptocur-rency ecosystem a breeding ground for questionable offerings. Stuck in the middle of this were banks and regulators wondering where they fit into the picture. 2017 was a year of precarity in the cryptocurrency ecosystem.

The small, grassroots group of early adopters were familiar with this uncertainty and learned to navigate it, making for a strong competitive advantage that kept the barrier to entry high for others; however, these shortcomings are unacceptable for a technology and asset class eyeing mainstream adoption.

Today, there is continual development aimed at making the cryptocurrency ecosystem less mysterious and more attractive to outside investors looking to dabble in the market. As cryptocurrency has become more legitimate and mainstream, so have the firms that operate in the space.

As standards have increased, it has become obvious which firms will make investments necessary to keep up, and which firms were sloppily thrown together to make a quick buck.

Fortunately, this gradual process to weed out bad actors has encouraged much needed development in crypto customer service, and developments made just months after the 2017 boom brought improvements in network latency, withdrawal and transfer times, and general operation and safety of wallets and exchanges.

These improvements have been one part self-imposed, and one part mandated by regulating authorities.

Cryptocurrency Markets & Regulation

As stated in the last chapter, government agencies such as the SEC, FTC, and CFTC are increasingly interested in understanding how to properly classify and regulate cryptocurrencies to reduce these risks, fraud, and illegal/unethical practices.

Regulating cryptocurrency is tough, though. While governments can demand some compliance from exchanges for operating in their country and servicing their citizens, once money is transferred to cryptocurrency, visibility reduces drastically. This is a natural byproduct of how cryptocurrency marketplaces were designed.

While some cryptocurrency purists dig their heels in at the site of any regulation, most investors, executives, and other actors have expressed a desire for more commonsense regulation in an effort to legitimize the space. While increased regulation reduces autonomy (which is always hard for cryptocurrency enthusiasts to give up), many doors are opened when government support is achieved.

Today, the relationship between regulators and cryptocurrency markets remains strained, but is improving. The challenge is to make everyone happy while regulating an asset that was created to be free from the grips of centralized authority and governance altogether.

Still, it's unclear how much each side is willing to give and take.

Fortunately, regulators, banks, investors and enterprises are collaborating to define and implement oversight that (almost) everyone can be happy with. The collaboration between these camps has bridged the previously feral cryptocurrency market-place to established institutions and facilitated noticeable improvements to customer service and offerings in the process.

Examples of the positive impact of recent regulation include:

- By becoming fully licensed and registered in the United States (or other nations), exchanges such as Gemini have gained the ability to trade fiat currency to cryptocurrency. This type of development has fundamentally altered how money enters and flows through the cryptocurrency ecosystem, since, as referenced in Chapter 4, times existed when there were very few options to safely, efficiently, and inexpensively purchase cryptocurrency.

- In 2018, Coinbase announced the launch of its own cryptocurrency index fund and there have been others working to follow suit. There are many altcoin-based hedge funds and investment offerings; however, most of them are not fully regulated and recognized by the SEC, meaning they may only trade on alternative exchanges where regulation is less strict. The creation of these funds directly followed the commencement of Bitcoin options and futures trading on the Chicago Board Options Exchange (CBOE) and the Chicago Mercantile Exchange (CME) in late 2017. This milestone marked the beginning of cryptocurrency derivatives trading outside dedicated cryptocurrency platforms. Today, many leading US exchanges and financial services firms offer cryptocurrency investing and derivatives trading.

- We are inching closer to recognized cryptocurrency ETFs and other large, managed investment vehicles, and firms have begun filing for patents and lobbying regulators to work towards the possibility. Regulatory agencies have made promising comments about rule changes, although they remain concerned over lack of controls in the cryptocurrency markets and potential for fraudulent activity—and for good reason.

These developments highlight how effective collaboration and regulation are helping the space progress. As the landscape matures in this way, we will continue to see the asset class shift, transform, and fall into place.

Whether we like it or not, regulation will likely continue to heavily impact how the asset is perceived, positioned, transferred, and used.

Cryptocurrency: Risks and Promise

Risks

The risks inherent to the cryptocurrency market largely stem from its questionable beginnings, immaturity, and lofty goals regarding its place in our financial system.

While marked improvements have been made to exchanges and wallets in general, a time existed (very recently) when the standard for security and navigability of these platforms was incredibly low, and extreme due diligence was required. Remnants of this disorder are still palpable today.

The decline and debrief period that followed the 2017 bull run revealed the precarity and volatility inherent to the marketplace, as well as the unscrupulous actors that perpetuated the impropriety.

Inability to remedy these flaws could lead to a permanently contaminated marketplace that will be deemed uninvestible by the market majority, due to its instability. In the absence of this stabilization, buy-in and support from regulating authorities and established entities is unlikely. As stated above, a strained relationship with these entities may prohibit the marketplace from experiencing advancements needed to keep it on a healthy path of development.

Promise

The 2017 cryptocurrency boom undoubtedly brought cryptocurrency markets—along with blockchain technology—to center stage, and this uptick in interest encouraged diverse investment interest and a relatively quick legitimization of the entire ecosystem.

Today, exchanges operate similar to other asset marketplaces, and the multitude of wallet options grants latitude to the investor in choosing their level of control and responsibility.

Additionally, balancing doses of regulation and competition have brought improvements to the customer service and general operation of most service providers. Customers now have access to a variety of reputable and reliable cryptocurrency exchanges and wallet services.

In addition, several brokerage services and mobile applications for retail investors are offering their clients the ability to invest in cryptocurrencies on a familiar interface.

On the funding side, ICO campaigns introduced a part-IPO, part-crowdsale model for raising funds from retail investors. This model proved unique by reducing the barriers for everyday investors wishing to invest in both an emerging market and early-stage startups building potentially cutting-edge technology, a privilege previously reserved for the likes of venture capital firms and angel investors.

Generally, it seems that cryptocurrency, as an ecosystem and investible asset, is increasingly embraced by the financial world, governments, and other established entities.

Final Thoughts and Further Reading

Cryptocurrencies and blockchain are here to stay.

Blockchain technology offers an improvement to extant systems that are bogged down by inefficiencies and over processing, while making a case for the sovereign management of data. Similarly, the rise of cryptocurrency has generated a unique asset class, while simultaneously offering a new method of storing and exchanging value that operates harmoniously with the modern digital economy.

While cryptocurrency and blockchain have followed separate

development paths, their persistence and interrelatedness cannot be ignored. 2017 & 2018, while tumultuous and frenzied, confirm that the asset class has generated sufficient support to justify continued interest and investment.

The informed consensus has begun morphing from "proceed with caution" to "ignore at your own risk," due to the recently realized potential of this new technology. This book endeavored to provide a comprehensive look at how the asset class came into existence, why it is important, and how to become equipped to invest in it prudently.

The topics covered in this book were very diverse and contain a variety of subtopics that evolve quickly by nature. I hope this book provided a broad foundation for this fascinating and relevant topic, and I hope you go on to consider sensible engagement with the marketplace or other research materials. For further reading, I endorse and appreciate many adjacent publications and their authors, such as those listed below, for their unique perspective and more focused viewpoint on certain branches of these topics:

- *The Bitcoin Standard* by Saifedean Ammous specifically focuses on Bitcoin as a store of value and the potential of this type of technology in transforming our financial system.

- *Blockchain Revolution* by Don and Alex Tapscott was originally written in 2016, but updated in October 2018. *Blockchain Revolution* provides an exciting view of blockchain's potential in the future of commerce from two thought leaders in the space.

- *Mastering Bitcoin* by Andreas Antonopoulos dives deep into the technical aspects of the blockchain. Readers unfamiliar with programming may not understand some concepts in the book, but it is probably still worth the read, since it provides a thorough understanding of distributed systems from a dedicated computer scientist and early thought leader.

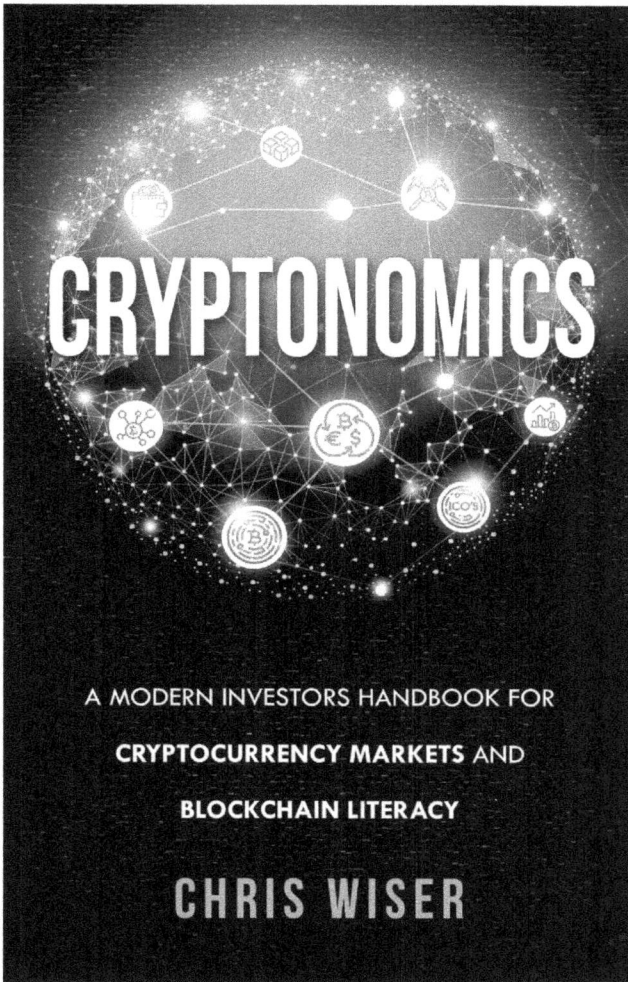

CRYPTONOMICS

A MODERN INVESTORS HANDBOOK FOR

CRYPTOCURRENCY MARKETS AND

BLOCKCHAIN LITERACY

CHRIS WISER

IF YOU ENJOYED this book and learned something valuable in the process, please think about leaving me an honest review on Amazon. Thank you!

Chris Wiser